Instructor's Guide

W9-AXX-008

Programmable Logic Controllers

Hardware and Programming

by
Max Rabiee
Associate Professor of Electrical and Computer Engineering Technology,
University of Cincinnati

Contributing Editor:
Stephen Fardo
Professor, Department of Technology, Eastern Kentucky University

Publisher
The Goodheart-Willcox Company, Inc.
Tinley Park, Illinois

Introduction

Programmable Logic Controllers—Hardware and Programming is an introductory text that explores many aspects of PLCs in an easy-to-understand manner. The key concepts of PLCs are discussed using a comprehensive approach to enhance learning. The text organization begins with basic concepts and progresses to system level applications. Applications, testing procedures, and operational aspects of PLC equipment and systems are discussed. This textbook emphasizes the PLC's practical use in industry.

Features of the Textbook

The *Programmable Logic Controllers—Hardware and Programming* textbook describes the most common programmable logic controller functions and provides examples using the Allen-Bradley Small Logic Controller (SLC 500) series programmable logic controllers. The textbook describes the PLCs and their use in process and industrial control systems, PLC theory, and PLC selection for various applications. PLC wiring and programming are covered with numerous examples.

The textbook has been divided into 15 chapters.

Chapter 1   Programmable Logic Controller (PLC) Overview

Chapter 2   PLC and Control System Components

Chapter 3   Number Systems and Codes

Chapter 4   Creating Relay Logic Diagrams

Chapter 5   PLC Programming

Chapter 6   Programming Logic Gate Functions in PLCs

Chapter 7   PLC Timer Functions

Chapter 8   PLC Counter Functions

Chapter 9   PLC Math Functions

Chapter 10   PLC Logic Functions

Chapter 11   PLC Compare, Jump, and MCR Functions

Chapter 12   PLC Subroutine Functions

Chapter 13   Sequencer Functions

Chapter 14   PLC Networks in Manufacturing

Chapter 15   Troubleshooting and Servicing the PLC System

Each chapter begins with an outline, Learning Objectives, and an introduction to the topic. Key words and terms in each chapter are highlighted in bold/italic type. Each chapter concludes with a summary of the important points, a glossary defining all of the key words and terms from the chapter, and Review Questions.

Features of the Laboratory Manual

The *Laboratory Manual* for *Programmable Logic Controllers—Hardware and Programming* is designed to supplement PLC training and works in conjunction with the textbook. Included are activities to lead students through a two-semester course in programmable logic controllers. The activities are written to give students hands-on experience practicing PLC programming. The diagrams and activities begin with basic concepts and progress to more complex applications. The activities and diagrams in the manual are formulated on the Allen-Bradley Small Logic Controller (SLC 500) series programmable logic controllers and Rockwell Automation's *RSLogix 500* software.

The laboratory activities can be divided into four categories based on their level of complexity.

▶ Laboratory activities 1 through 17 are basic or beginning level PLC assignments.

▶ Laboratory activities 18 through 33 are intermediate level PLC assignments.

▶ Laboratory activities 34 through 41 are advanced level PLC assignments.

▶ Laboratory activities 42, 43, and 44 are PLC networking lab assignments.

Lab Assignment 2 asks questions that cover materials found in Chapters 1, 2, 4, and 5. The other activities correlate to the textbook chapters as follows.

Chapter 3:   Lab Assignment 1

Chapter 5:   Lab Assignments 3, 4, 5, 6, 7, 8, 9, 10, and 11

Chapter 6:   Lab Assignments 12, 13, 14, 15, 16, and 17

Chapter 7:   Lab Assignments 18, 19, and 20

Chapter 8:   Lab Assignments 21, 22, 23, 24, 25, 26, 27, and 28

Chapter 9:   Lab Assignments 29, 30, 31, 32, and 33

Chapter 10:   Lab Assignment 34

Chapter 11:   Lab Assignments 35 and 36

Chapter 12:   Lab Assignment 37

Chapter 13:   Lab Assignments 38, 39, 40, and 41

Chapter 14:   Lab Assignments 42, 43, and 44

Materials and Equipment

The following list contains all the equipment and materials required to perform the activities in the *Laboratory Manual*. In some activities, different components can be substituted for the components listed here.

▶ Allen-Bradley fixed SLC 500 PLC with 1747-L20A processor or equivalent (2)

▶ Allen-Bradley SLC 503 PLC or equivalent

▶ Split-phase ac induction motors (2) (fractional horsepower)

▶ Single-pole, single-throw (SPST) switches (6)

▶ Normally open pushbuttons (3)

▶ Normally closed pushbutton

▶ Red, green, and white pilot lights

▶ Bell (6-volt)

▶ Scientific calculator

In addition to the hardware listed, these activities are designed around Rockwell Automation's *RSLogix 500* PLC programming software, the most popular software for programming the Allen-Bradley SLC 500 series PLC.

Software

Included with the *Laboratory Manual* is the *LogixPro* PLC simulation software. *LogixPro* is a tool to facilitate student learning of the fundamentals of RSLogix ladder logic programming. *LogixPro* will allow students to practice and develop their programming skills at home as well as in the classroom. Note that *LogixPro* is not a replacement for RSLogix. There is no support for file exchange, nor is there support for communication with actual Allen-Bradley products. *LogixPro*, instead, provides a complete software based simulation solution. It is a solution designed specifically for training.

An inclusion with the *LogixPro* software is the *ProSim-II Simulation* package that graphically simulates process equipment, such as batch mixing systems, traffic lights, and garage doors. The *ProSim-II Simulation* package gives students the synchronous and interactive experience of real industrial processes.

Using LogixPro Software

The first time the *LogixPro* CD is placed in the disc drive, the installation program should run automatically. To complete the installation, the students should simply follow the prompts provided. This will install the *LogixPro* software. If the installation does not run automatically, the installation program can be started manually. Use Windows Explorer to locate the file titled **LogixCD.exe** on the CD. This file is located in the root directory of the CD. Double clicking this file will start the installation.

In order to run *LogixPro*, the CD must be in the computer's disc drive and the drive door must be closed. *LogixPro* can then be started using the icon located on the Windows desktop or the icon located at **Start | Programs | TheLearningPit | LogixPro**.

The CD also comes with the program *PSIM*, an emulator for Allen-Bradley PLC/2 family PLCs. Double clicking on **simSetup.exe** found in the **PSIM** folder of the CD will invoke the self-extracting installation program. It will lead you through the installation and setup of *PSIM*.

For additional information regarding *LogixPro*, visit http://TheLearningPit.com.

Table of Contents

Advanced Level PLC Assignments

PLC Networking Lab Assignments

Answers to the Textbook

Chapter 1

Answers to Review Questions

1. Microprocessor unit (MPU)
   Read only memory (ROM)
   Random access memory (RAM)
   Decoder circuit
   Clock circuit
   Peripheral interface adaptor
   Input buffer
   Output buffer

2. A discrete input port is either open (logic low) or closed (logic high). A variable input port is capable of converting variable voltages into binary data.

3. Central processing unit (CPU)
   Power supply
   Input module(s)
   Output module(s)

4. Static RAM (SRAM)
   Dynamic RAM (DRAM)

5. Masked (or preprogrammed) ROM
   Programmable ROM (PROM)
   Erasable programmable (EPROM) or ultraviolet-erasable programmable EPROM (UVEPROM)
   Electrically erasable PROM (EEPROM) or flash ROM

6. The address decoder circuit ensures that only one device is communicating with the microprocessor at any given moment.

7. With a fixed PLC system, the power supply, CPU, input, and output groups are all in the same enclosure. With a modular PLC, each component is in a separate expansion slot.

8. The battery backup ensures that program and data files on the system RAM are retained in case of power failure.

Chapter 2

Answers to Review Questions

1. Channel zero is used for connecting bar code readers or printers to the PLC. Channel one is usually utilized for downloading or monitoring the PLC program.

2. Data highway communication connects PLC devices together in a networked environment.

3. System PLC memory stores information needed to carry out the user program. The user memory holds the ladder logic diagram.

4. Input status
   Output status
   Timer status

Counter status

Other functions (sections used for addition, subtraction, multiplication, division, sequencer, shift registers, and comparison)

5. Seven factors are listed in the text.
   Manufacturer's support
   Serviceability
   Flexibility
   Expandability
   Programming software
   Training
   Documentation

6. Work is done when an applied force moves an object. Torque is a force that causes an object to rotate.

7. Magnetic flux is the magnetic lines of force that flow from the north pole to the south pole. It is measured in webers (Wb). Magnetic flux density is the number of lines of flux per unit of surface area. It is measured in webers per square meter or teslas (T).

8. Faraday's law states that when a conductor moves in a magnetic field such that the flux lines are cut, a voltage is induced in that conductor. $e = B \times L \times N \times \sin(\theta)$.

9. A continuous duty motor is designed to operate continuously at its rated speed. An intermittent duty motor is designed to operate for a short period of time every so often.

10. Fractional horsepower
    Integral horsepower

11. Speed regulation is the ability of a motor to maintain its rated speed as the load on the rotor shaft is changed.

12. Separately excited dc motor
    Series-connected dc motor
    Shunt-connected dc motor
    Compound dc motor

13. Squirrel cage induction motor. Rotor bars are embedded into rotor slots and connected together by two end rings.
    Wound rotor induction motor. Rotor has winding coils.

14. Slip is the difference between the rotor speed and stator field speed.

Chapter 3

Answers to Review Questions

1. 4-bit   1111 binary   =   15
   8-bit   1111 1111 binary   =   255
   16-bit   1111 1111 1111 1111 binary   =   65535
   32-bit   =   4294967295
   64-bit   =   18446744073709551615

2. 1101.11   =   13.75
   10011101.01   =   157.25
   101101100101.101   =   2917.625
   1010110011101110.001   =   44270.125

3. 2.5   =   10.1
   3.625   =   11.101
   65   =   1000001
   25.125   =   11001.001
   217.00125   =   11011001.00000000011 (approx.)

4. $1100 + 1111$ $= 11011$
   $11000111 + 11000000 = 110000111$
   $10101011 + 11110011 = 110011110$
5. $1100 - 1001$ $= 11$
   $10001100 - 10000101 = 111$
   $11100011 - 10101010 = 111001$
6. 10101×1011 $= 11100111$
   $11011100 \times 10101111 = 1001011001100100$
7. $11000111 \div 1010 = 10011$
   $10101111 \div 1111 = 1011$
8. 43794
9. 1010111001101.1010001
10. 37.38 (approx.)
11. $25 = 31_O$
    $127 = 177_O$
    $365 = 555_O$
12. $110111 = 67_O$
    $1111000111.01 = 1707.2_O$
    $1000111001.1 = 1071.4_O$
    $1110011100.010 = 1634.3_O$
13. $111 = 7 = 0111_{BCD}$
    $1110 = 14 = 0001\ 0100_{BCD}$
    $10000 = 16 = 0001\ 0110_{BCD}$
    $100010 = 34 = 0011\ 0100_{BCD}$
    $100110 = 38 = 0011\ 1000_{BCD}$
14. Student answers will vary. Check Table 3-2.
15. Student answers will vary. Check Table 3-2.

Chapter 4

Answers to Review Questions

1. You can place as many input instructions as desired in series or in parallel.
2. You can only place one output instruction per rung, so you cannot place any in series.
3. Output instruction
4. You can use the same input instructions more than once. Output instructions can be used only once.
5. A momentary pushbutton closed or opened (has its state changed) only while it is being pressed. A locked-position pushbutton opens or closes (changes state) when pressed and remains in the changed state until it is pressed again.
6. Normally open switches are always open, until they are forced to close. Normally closed switches are always closed, until they are forced to open.
7. Limit switch—detects when an object contacts the handle of the switch.
   Proximity switch—detects when an object breaks a light beam. Detects a part without coming into contact with the part.
   Temperature switch—detects when a temperature rises above or goes below a set point.
   Level switch—detects when a liquid level rises above or falls below a certain point.
   Flow meter switch—detects when a preset speed of flowing liquid or gas is reached.
   Pressure switch—detects when a preset pressure is reached.

8. Dashed lines indicate that the input devices operate simultaneously.

9. Relay coils are energized to cause normally open contacts to close or normally closed contacts to open.

10. Two types of overload relays are available: temperature (or thermal) overload and magnetic overload. Temperature overload relays detect overcurrent through excessive temperature rise generated by the line current. Magnetic overload relays detect overcurrent using the magnetization the line current generates.

11. In the jog mode, the motor operates only when the jog pushbutton is pressed.

Chapter 5

Answers to Review Questions

1. 1747-L20A

2. 1746-IA16

3. 1746-OA16

4. Output instruction

5. No. Output address instructions can be used only once.

6. Yes. Input address instructions can be used more than once.

7. Module zero

8. File 2

9. B3/0 through B3/255; B10/0 through B10/255

10. Left click on the instruction to select it.
    Right click to open its shortcut commands.
    Select the Forced On command or Forced Off command.
    Then select the Enable Force command.

11. No. You can generate a dangerous environment by turning on input/output devices via computer keystrokes.

12. Status (S2)

13. Left click to select the file.
    Right click to open the shortcut menu.
    Select (Copy/Rename/Open/Save).

14. XIO—Examine if open: PLC instruction for the normally open input device.
    XIC—Examine if closed: PLC instruction for the normally closed input device.

15. See Figure 5-34.
    General: Title Page, Processor Information, I/O Configuration, Custom Data Monitor, Cross Reference, and Multipoint List
    Data Base: Address/Symbols, Information Comments, Symbol Groups
    Program Files: Program File List, Program Files
    Data Files: Data File List, Data Files, Memory Usage

Chapter 6

Answers to Review Questions

1.

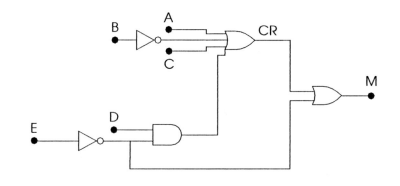

2.

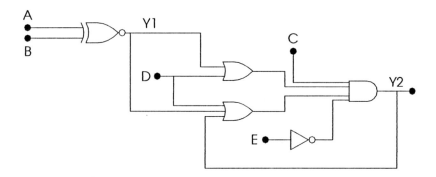

3.

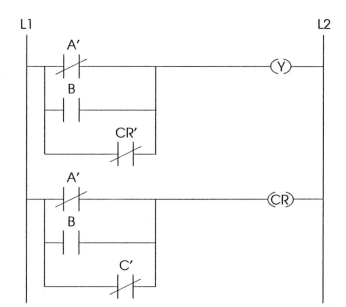

4.

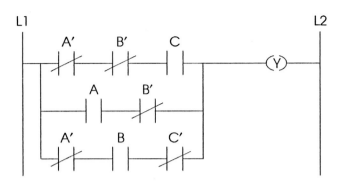

5.

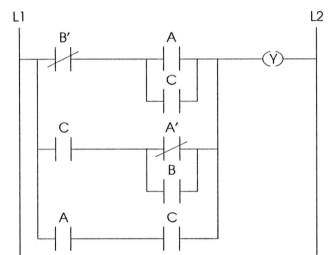

6.

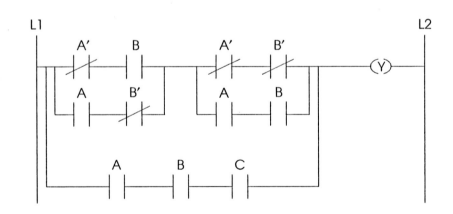

7. Boolean expression: Y = AC + BC

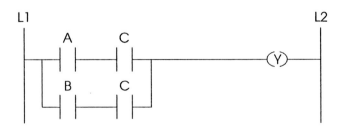

8. Boolean expression: Y = A'BC + AB'C + ABC'

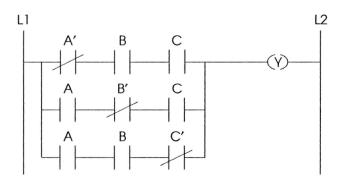

9. Boolean expression: (B'D'E'F' + AD'E'F' + CD'E'F') + (G'H'J + G'HJ) = Y

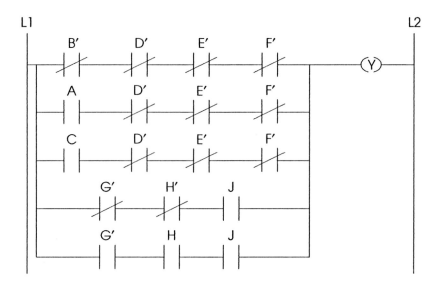

10. Boolean expression: Y = A'BE'Y + C'E'Y + D'E'Y

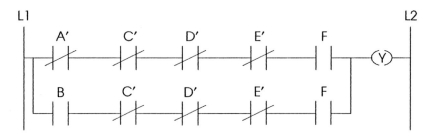

Chapter 7

Answers to Review Questions

1. The timer base value is found by multiplying the timer base number by the preset value of a PLC.
2. A retentive timer will hold its accumulated value when it is de-energized. A nonretentive timer resets when it is de-energized.
3. Timer ON-delay functions must be energized to start operation. Timer OFF-delay functions must be de-energized to start operation.
4. The Allen-Bradley SLC 500 has 256 timers with data files T4:0 through T4:255.
   Accumulated timer registers: T4:0.ACC through T4:255
   Preset timer registers: T4:0.PRE through T4:255.PRE

5. An ON-delay function is started by energizing it.
6. An OFF-delay function is started by de-energizing it.
7. The reset (RES) function must be used to reset a retentive timer.
8. The timer done status bit in an ON-delay function is energized when the timer is done.
9. The timer timing status bit in an ON-delay function is energized when the timer is timing.
10. The timer done status bit in an OFF-delay timer function is de-energized when the timer is done.

Chapter 8

Answers to Review Questions

1. Two. Count up and count down
2. In the count up function, the accumulated register increments whenever the counter input device changes state. It is incremented with a low-to-high pulse.
3. The reset (RES) function is used to reset all counters.
4. The done bit is de-energized.
5. The done bit is energized.
6. The Allen-Bradley SLC 500 PLC has 256 counters with data files C5:0 through C5:255. The accumulated registers are addressed C5:0.ACC through C5:255.ACC.
7. The preset counter registers are addressed C5:0.PRE through C5:255.PRE

Chapter 9

Answers to Review Questions

1. Three. Source A, source B, and the destination.
2. Yes. The destination register holds the sum.
3. No. At least one source must be a register.
4. Yes. Both sources are allowed to be registers.
5. No. At least one source must be a register.
6. Yes. The divide function carries out only integer point division.
7. The dividend is placed in source A. The divisor is placed in source B.
8. The remainder can be found using the subtract and divide functions: Remainder = Dividend − (Quotient × Divisor)
9. The largest number must be held in a 16-bit destination register. Therefore, the number cannot be larger than $FFFF_h$ (65,536 decimal).
10. No. The destination holds the result of the subtract function. Therefore, it must be a register.

Chapter 10

Answers to Review Questions

1. There are 256 control registers. R6:0 through R6:255.
2. A bit shift left function must have a designated control register. If R6:0 is the control register, then the R6:0/DN bit is energized when the BSL function is done.
3. B3/0 through B3/255 or B10/0 through B10/255 can be used for BSR functions.
4. The bit where the last shifted bit exits is called the bit address. The bit address exits and reenters at the bit at the beginning of the length. (See Figure 10-8.)

5. Source A and source B in an OR logic function can be 16-bit numbers. Therefore, 16 bits are ORed in an OR logic function.

6. $A' + B' = (A \times B)'$

7. $A' \times B' = (A + B)'$

8. The control bit is energized when the BSR function is on.

9. $(A' \times B') + (A \times B) = (A \oplus B)'$

10. A low-to-high input pulse will cause the bit shift right function to shift the bits in its data file once to the right.

Chapter 11

Answers to Review Questions

1. The six compare functions are: equal to (EQU), not equal to (NEQ), less than (LES), greater than (GRT), less than or equal to (LEQ), and greater than or equal to (GEQ). For examples, see Figures 11-2 through 11-7.

2.

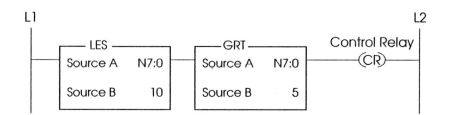

3.

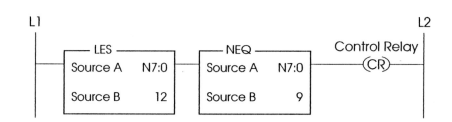

4. The JMP function is energized to activate it.

5. The MCR function is de-energized to activate it.

6. The state of a jumped instruction remains the same.

7. Instructions between MCR rungs are turned off.

8. The MCR function is used to shut down or turn off the PLC system when there is a power outage to the control system.

9. Yes

10. Yes. Student examples will vary.

11. Yes. Student examples with vary.

12. No

13. Yes

14. Yes

Chapter 12

Answers to Review Questions

1. File number two (LAD 2)

2. Theoretically, 253 subroutine files can be created. They are file #3 through file #255 or LAD 3 through LAD 255.

3. Subroutine (SBR)

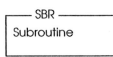

4. Jump to subroutine (JSR)

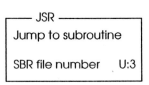

5. Jump to subroutine (JSR): Used to call a subroutine from the main program file.
   Return (RET): Used to return the program counter to the main file from a subroutine file.

Chapter 13

Answers to Review Questions

1. The control registers are R6:0 through R6:255.

2. Field in a sequencer function that holds a hex number that will be added to the 16-bit word in a data file in order to hide (mask) some of the bits.

3. The area that holds the total number of steps that must be completed before the sequencer done status bit coil is energized.

4. The field that indicates the step that is desired to start the sequencer function.

5. A low-to-high pulse on the step input will advance the sequencer position.

6. Three: Sequencer output function (SQO), sequencer compare function (SQC), and sequencer load function (SQL)

7. The done bit is energized when the sequencer has completed the steps. The enable bit is energized when the sequencer is on.

8. The found bit is energized when the data on the input port matches the data stored in the sequencer data file.

9. The reset (RES) function is used to reset the sequencer.

10. B3, B10, and N7

11. Cascading sequencer functions will increase the number of steps.

12. Parallel sequencer connections will increase the number of sequencer outputs.

Chapter 14

Answers to Review Questions

1. 16 Mbs and 100 Mbs

2. Baseband—Special 50-ohm cable that carries digital signals only. Also called Ethernet.
   Broadband—Cable that carries multiple analog signals.
   Carrier band—Cable that carries a single analog signal.

3. 2 Gbs

4. Ring topology—Topology that transmits a signal through a closed loop and the signal is copied by the intended destination network node. The signal is absorbed by the original station that transmitted the signal.
   Bus topology—Topology that transmits a signal that is copied by the target station and the signal is then absorbed by terminating point resistors.
   Tree topology—Topology that has multiple branches that are connected together at one end.
   Star topology—Topology that has every node connected directly to the central node. The nodes have no connections to other nodes. The transmission of data by any station in the network is controlled by the central hub.

5. Peer-to-peer—Network that allows the master station to initiate communication by sending signals to the slave stations.
   Token ring—Network that uses a circulating token in the circuit. Network stations must hold the token in order to have access to the network medium for a predetermined amount of time.
   Carrier sense multiple access with collision detection (CSMA/CD)—Network scheme that allows each network station to listen to the communication line and try to get access to the transmission line.

6. Unshielded twisted pair (UTP) wires—Wire that does not have a mesh wire jacket. Each pair of two twisted wires is used for one communication channel.
   Shielded twisted pair (STP) wire—Wire that has a mesh wire jacket that is connected to earth ground at both ends of the medium. STP wires provide more immunity against electro-magnetic/electrical noise than the UTP wires.

Chapter 15

Answers to Review Questions

1. Five. Power, PC run, CPU fault, Forced input/output, Battery low

2. Forced mode is used for checking input/output ports and for troubleshooting PLC ladder logic diagrams.

3. Connect to the PLC and place the system in the run/monitor mode. Check S2 to open the status bit file dialog box to find the error flags.

4. Check hardware connections. Next, debug the ladder logic software. Finally, open S2 error file and clear the error bit.

5. The power status indicator light indicates that power has been applied to the PLC and the processor is energized.

6. The battery low indicator light turns on to indicate when the CMOS battery on the CPU board is running low.

Answers to the Laboratory Manual

Lab Assignment 1

1. c
   d
   b
   a

2. a. 7
      11
      42
      2494
   b. 1110
      101010
      10100010
      11001111
      111001011
   c. 1100.1
      101000011.101
      11010001110.0000001
      10011010100.0000000001
      111010001.00000000001
   d. 0101 0110
      0011 1000 0001
      0001 0001 0010 0001
      0100 0101 1000 0011
      0110 0110 1000 0101
   e. 63
      6376
      8941
      5085
      3327
   f. 7
      17
      130
      507
      2141

   g. 6
      29
      63
      129
      2356
   h. 8
      E
      4F
      22E
      4DB
   i. 17
      60
      674
      3023
      6988
   j. FE
      23
      9BE
      C21
      F8AB

3. 155
   101
   219
   85

4. 51.5
   97.3
   1871.625
   234.125

5. 1001110.101
   1110000.01
   101010111100.0000111
   1100000.11
   111110001.001101

Lab Assignment 2

1. Twelve input ports; Eight output ports

2. Central processing unit (CPU)
   Input module(s)
   Output module(s)
   Power supply

3. Read input ports
   Carry out (execute) the ladder logic instructions
   Update (write to) output ports

4. To protect the CPU from overvoltage (voltage spike) input signals from input ports. It also protects from short circuits on the output ports.

5. a. Status of the input ports saved on the PLC memory.
   b. Status of the output ports saved on the PLC memory.
   c. Includes the microprocessor, memory, and support chip in a PLC system.
   d. Hold groups of input and output ports. Also called I/O slots or I/O expansion slots.

6. a. Data file used for internal control relay contacts.
   b. Data files utilized for internal control relay contacts.
   c. File that holds the content of the control registers.
   d. File that holds the content of the counter functions.
   e. Data files used to hold the integer numbers.
   f. File that holds the PLC error flag(s).
   g. File that holds the content of the timer functions.

Lab Assignment 3

Activity 2.

Step 1. When normally open pushbutton one (PB1) and pushbutton two (PB2) are closed, the motor turns on. When PB1, PB2, and limit switch (LS1) are closed, both the red pilot light and the motor turn on.

Step 3.

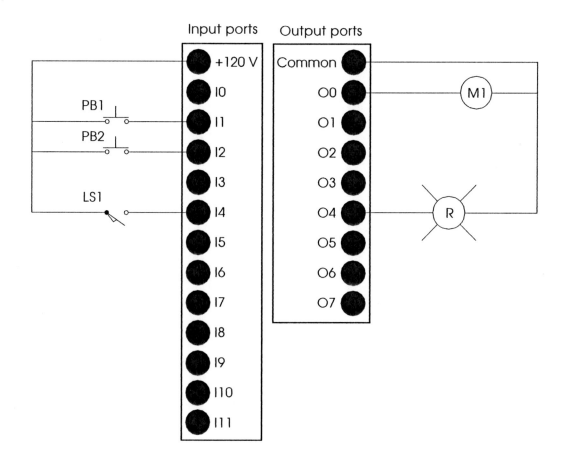

Step 4.

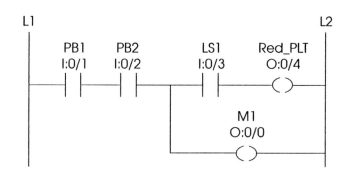

Lab Assignment 4

Activity 1.

Step 1. Normally closed pushbutton (labeled Stop) is emergency pushbutton to stop the system. When normally open switch (SW1) is closed, the motor (M1) turns on. If the normally open limit switch (LS1) is also closed, the red pilot light turns on and the motor turns off.

Step 3.

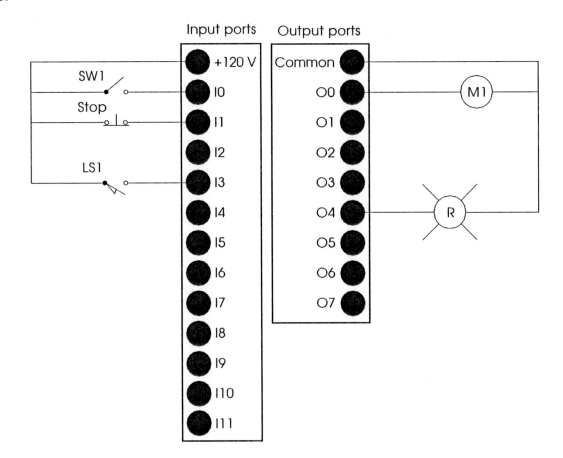

Step 4.

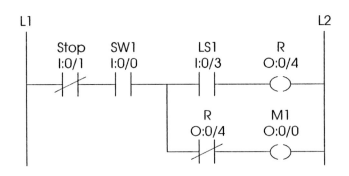

Activity 2.

Step 1. Normally closed pushbutton (PB) is the stop pushbutton for motor (M1). The green pilot light is on when the PLC system is turned on. Closing the normally open thermostat (TSW) will turn on the motor and the red pilot light. The motor will lock or seal itself and continue operating until PB is pressed. When the motor is on, the green pilot light turns off.

Step 3.

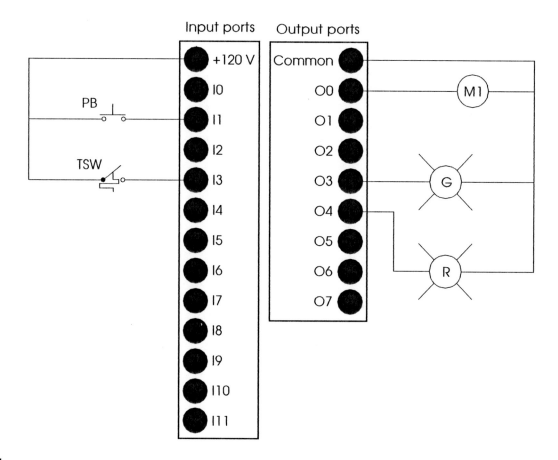

Step 4.

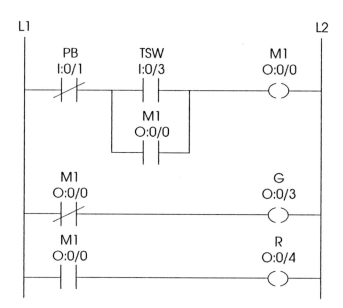

Lab Assignment 5

Activity 1.

Step 1. Normally closed pushbutton (Stop) is the emergency stop pushbutton. When either normally open limit switches one (LS1) and two (LS2), or limit switch three (LS3) are closed, control relay coil (CR) is energized. Closing switch one (SW1) will turn the motor (M) on.

Step 3.

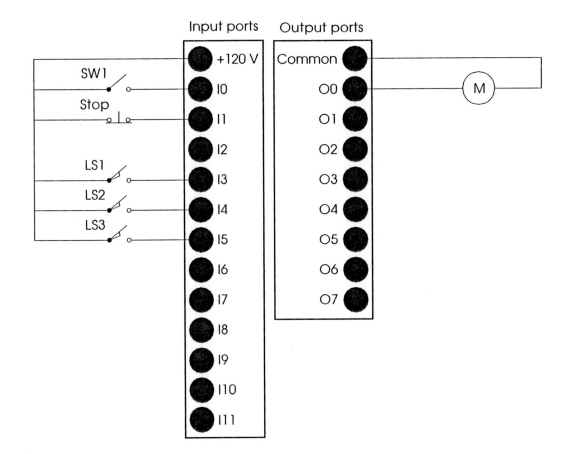

Step 4.

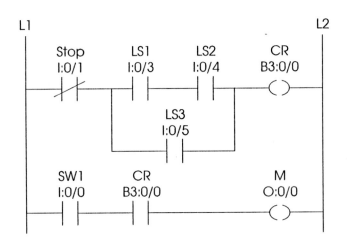

Activity 2.

Step 1. Closing the normally open switch (SW1) will start the motor (M) and turns on the red light (R). Opening the normally closed limit switch one (LS1) will turn the motor and red light off. Closing the normally open limit switch (LS2) will also turn off the motor and red light. However, it will turn the bell on. Press the stop pushbutton to turn all the outputs off.

Step 3.

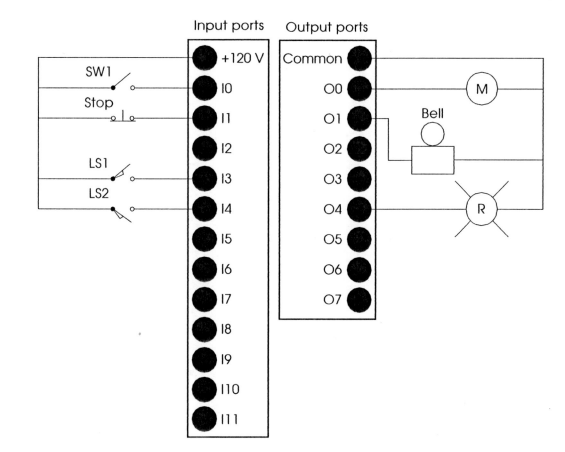

Step 4.

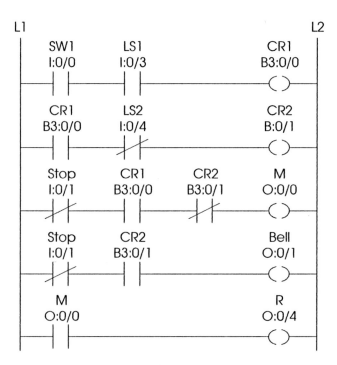

Lab Assignment 6

Activity 1.

Step 1. Press pushbutton one (SW1) to turn motor one (M1) on. It will lock itself in continuous run mode. Press pushbutton two (SW2) to turn motor two (M2) on. It will lock itself in continuous run mode. Closing limit switch one (LS1) or pressing the Stop switch will turn both motors off.

Step 3.

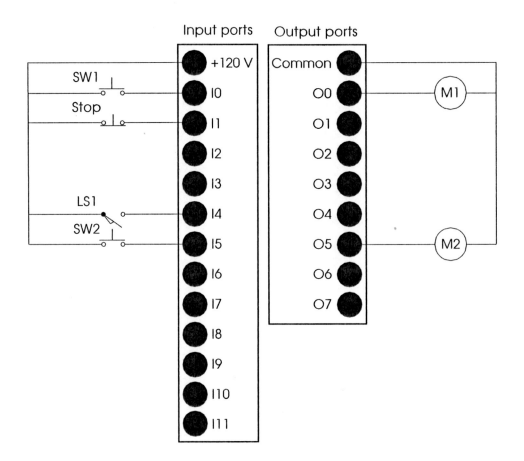

Step 4.

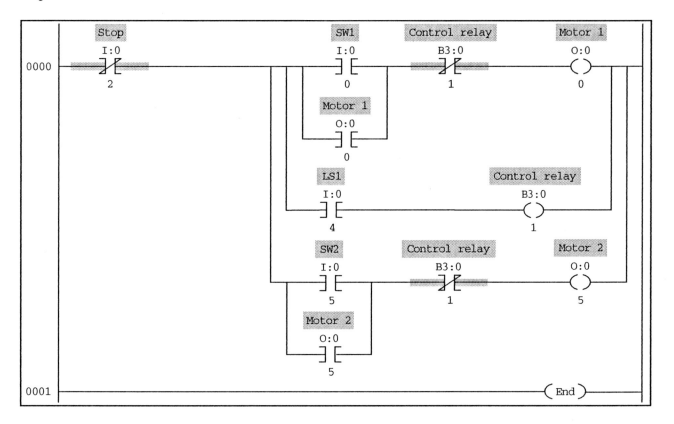

Activity 2.

Step 1. Press the green pushbutton (PBG) only to start motor one (M1). It will lock itself in the continuous run mode and the green pilot light turns on. Press the red pushbutton (PBR) to start motor two (M2). It will lock itself in the continuous run mode and the red light turns on. Pressing both pushbuttons will turn both motors on and only the white light turns on. Opening the temperature switch or pressing the stop pushbutton will turn every output off.

Step 3.

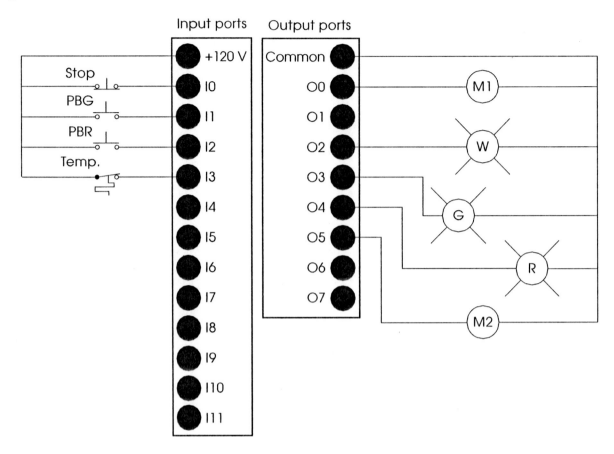

Step 4.

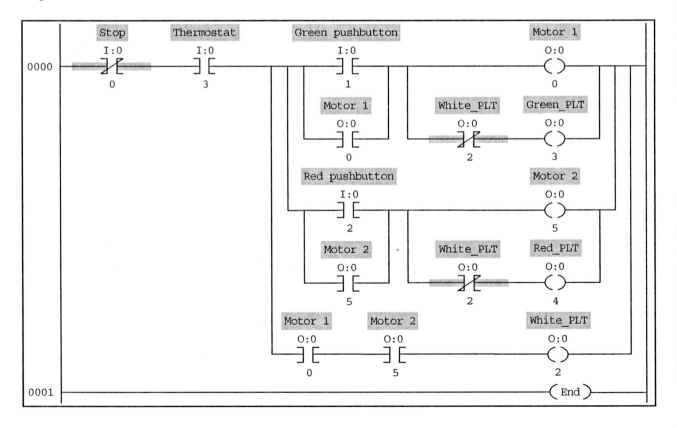

Lab Assignment 7

Activity 1.

Step 1. Close the switch (SW) to latch the contact in rung three and turn on the white pilot light.
   Press the red pushbutton to unlatch the contact in rung three and turn off the white pilot light.

Step 3.

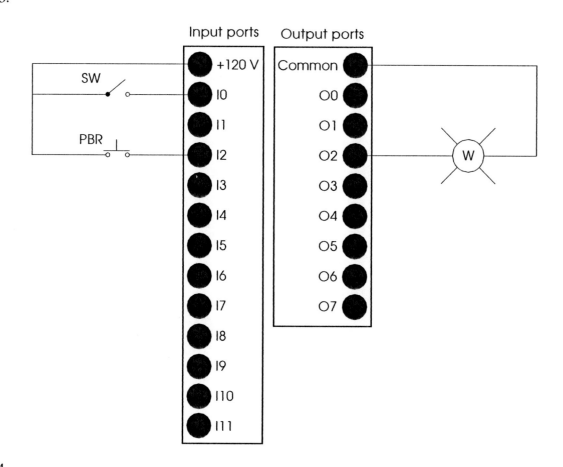

Step 4.

Activity 2.

Step 1. Press the black switch (SWB) to latch contact one (Latch 1) and turn on the green pilot light. Press the green pushbutton to unlatch contact one and turn off the green pilot light. Press the green pushbutton to unlatch contact one and turn off the green pilot light. The temperature switch (TSW) and red pushbutton (PBR) are used to latch and unlatch contact two in order to turn the red pilot light on and off.

Step 3.

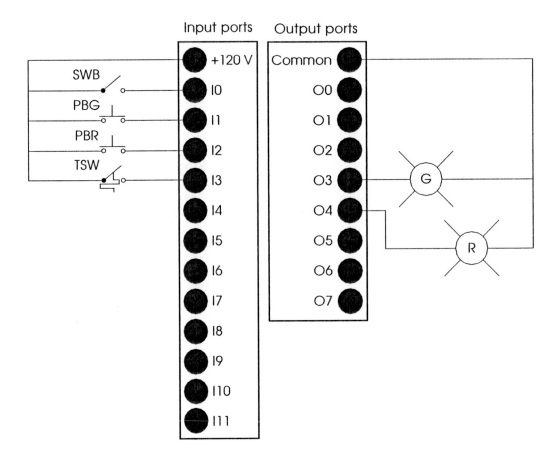

Step 4.

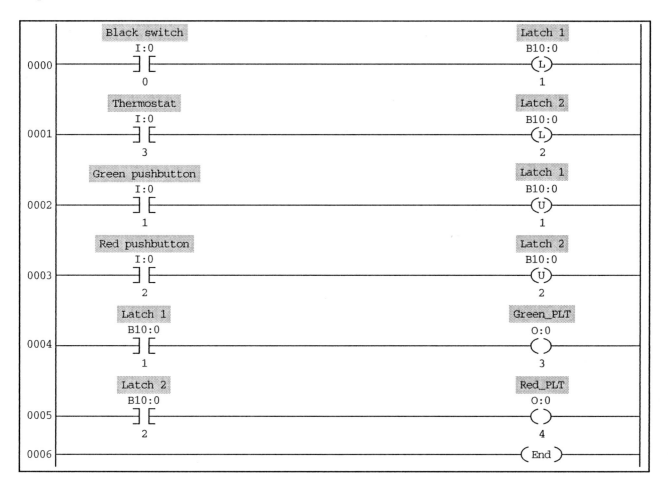

Lab Assignment 8

Activity 1.

Step 1. See PLC report in Step 4.

Step 2.

Input Device	PLC Input Port Address
Switch 1 (SW1)	I:0/4
Switch 2 (SW2)	I:0/5
Output Device	PLC Output Port Address
White PLT	O:0/2
Green PLT	O:0/3
Red PLT	O:0/4

Step 3.

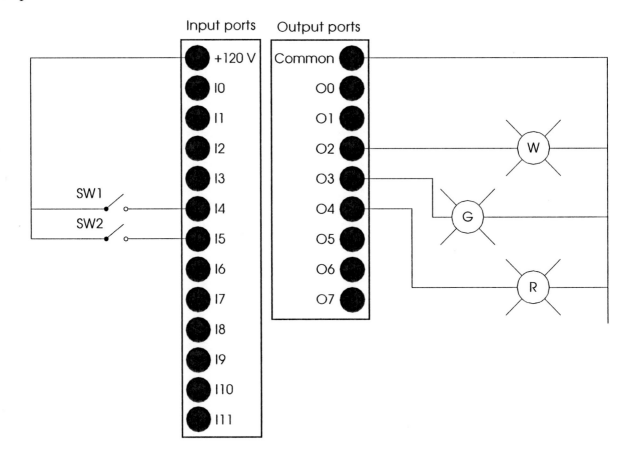

Step 4.

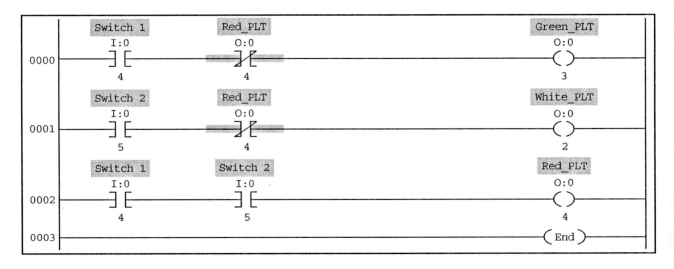

Activity 2.

Step 1. See PLC report in Step 4.

Step 2.

Input Device	PLC Input Port Address
Black switch	I:0/0
Green pushbutton	I:0/1
Red pushbutton	I:0/2
Output Device	PLC Output Port Address
Motor 1	O:0/0
Motor 2	O:0/5
White PLT	O:0/2
Green PLT	O:0/3
Red PLT	O:0/4

Step 3.

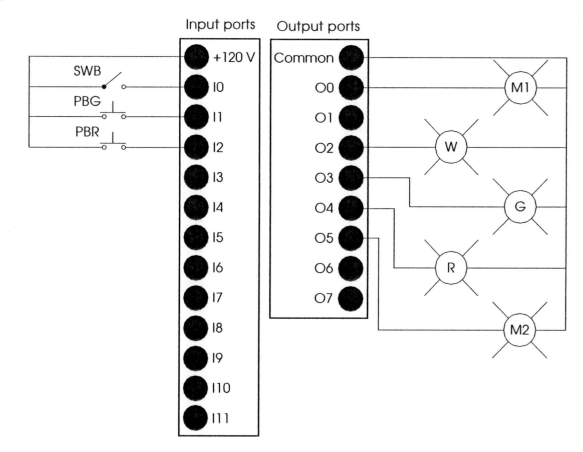

Step 4.

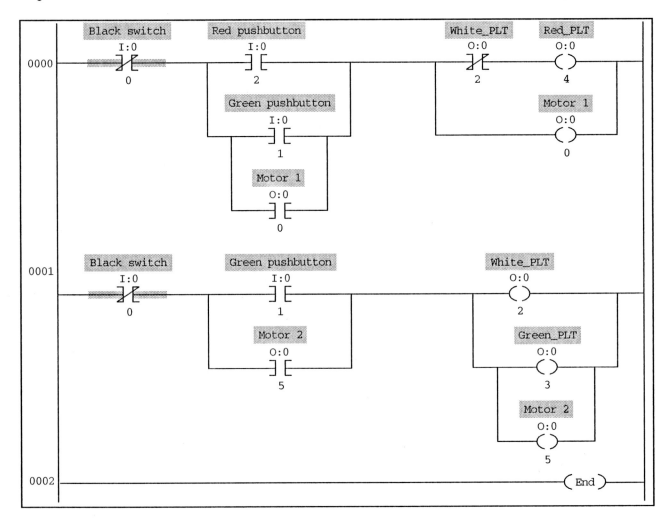

Lab Assignment 9

Activity 1.

Step 1. See PLC report in Step 4.

Step 2. Input Device	PLC Input Port Address
Black switch	I:0/0
Red pushbutton	I:0/2
Output Device	PLC Output Port Address
Motor 1	O:0/0
Motor 2	O:0/5
White PLT	O:0/2
Green PLT	O:0/3

Step 3.

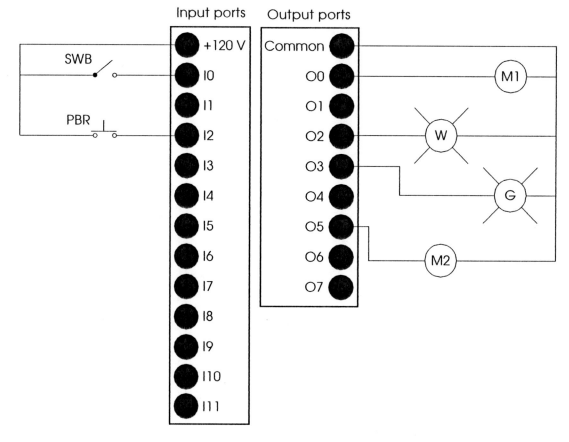

Step 4.

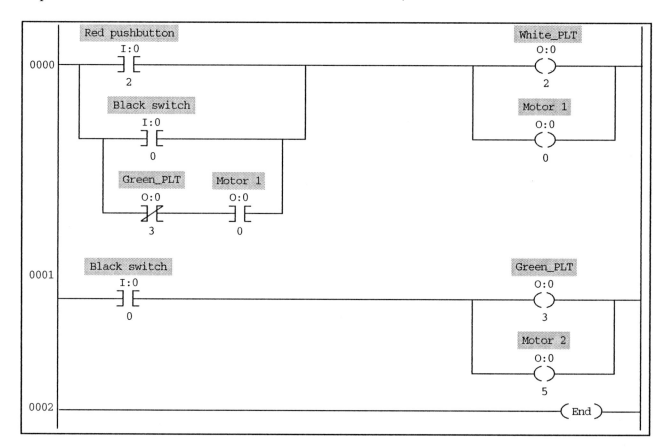

Activity 2.

Step 1. See PLC report in Step 4.

Step 2.

Input Device	PLC Input Port Address
Black switch	I:0/0
Green pushbutton	I:0/1
Red pushbutton	I:0/2
Thermostat	I:0/3
Output Device	PLC Output Port Address
Motor 1	O:0/0
Motor 2	O:0/5
White PLT	O:0/2

Step 3.

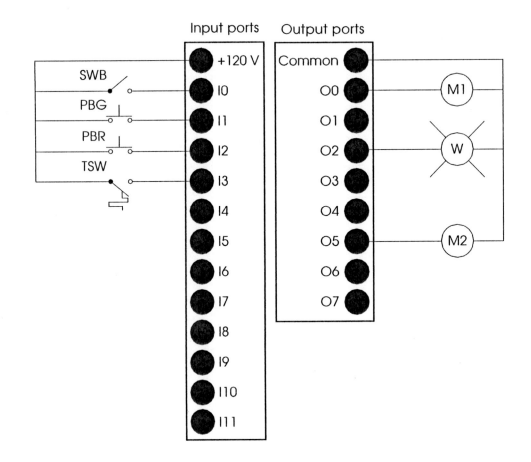

Step 4.

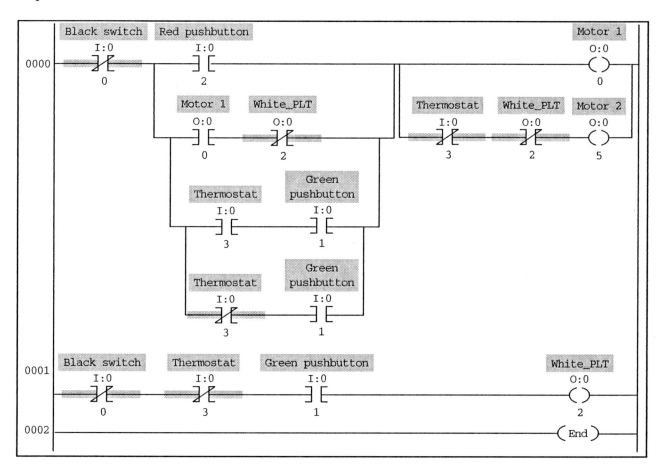

Lab Assignment 10

Activity 1.

Step 2.

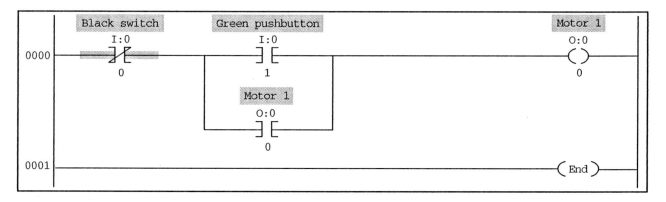

Activity 2.

Step 2.

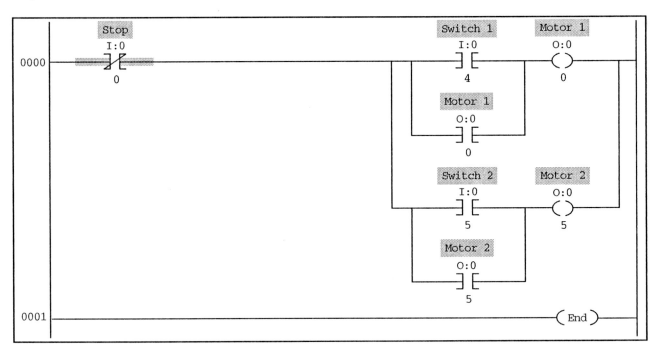

Lab Assignment 11

Activity 1.

Step 2.

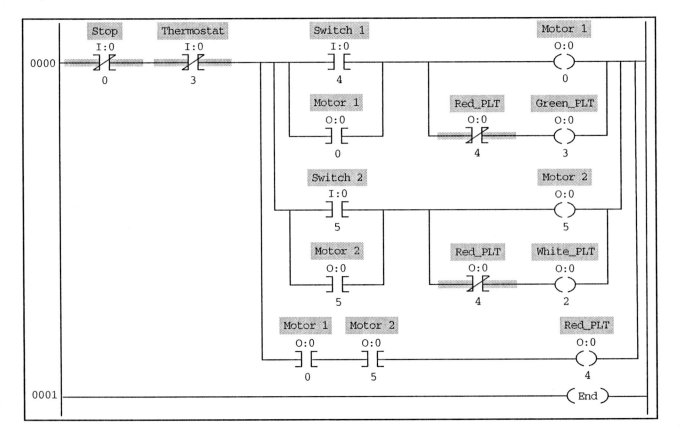

Activity 2.

Step 2.

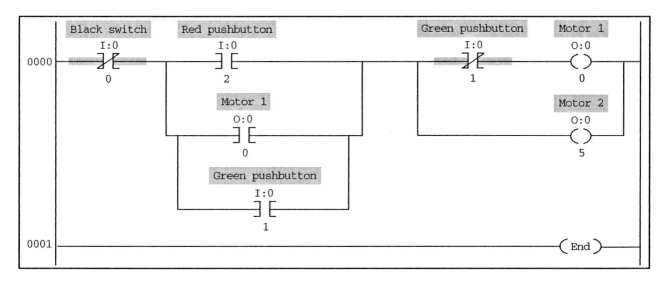

Lab Assignment 12

Activity 1.

Step 1.

<u>AND gate</u>

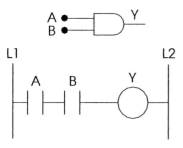

A	B	Y
0	0	0
0	1	0
1	0	0
1	1	1

<u>OR gate</u>

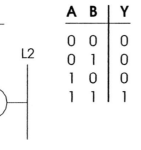

A	B	Y
0	0	0
0	1	1
1	0	1
1	1	1

<u>NAND gate</u>

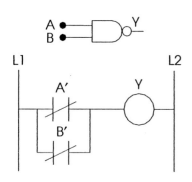

A	B	Y
0	0	1
0	1	1
1	0	1
1	1	0

<u>NOR gate</u>

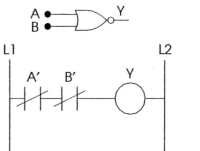

A	B	Y
0	0	1
0	1	0
1	0	0
1	1	0

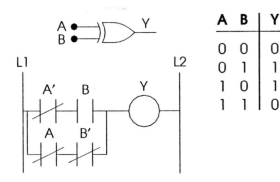

XOR gate

A	B	Y
0	0	0
0	1	1
1	0	1
1	1	0

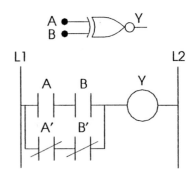

XNOR gate

A	B	Y
0	0	1
0	1	0
1	0	0
1	1	1

Step 3. See ladder diagrams in Step 1.

Activity 2.

Step 1.

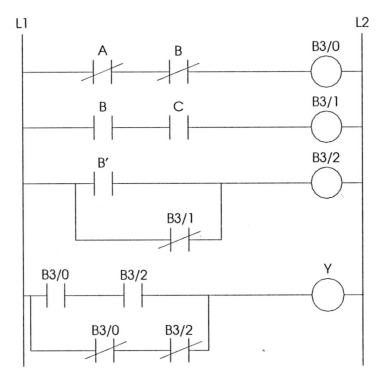

Internal bit B3/0 = CR1   Internal bit B3/1 = CR2   Internal bit B3/2 = CR3

Step 3. See ladder diagram in Step 1.

Lab Assignment 13

Activity 1.

Step 1. Y1 = BC + AC + AB + ABC = AB + AC + BC

Y2 = A'B'C + A'BC' + AB'C' = A'(B'C + BC') + AB'C' = A' (B $\oplus$ C) + AB'C'

Y3 = A'B'C' + A'B'C + A'BC' + AB'C' = B'C' + A'(B'C + BC') = B'C' + A'(B $\oplus$ C)

Step 3.

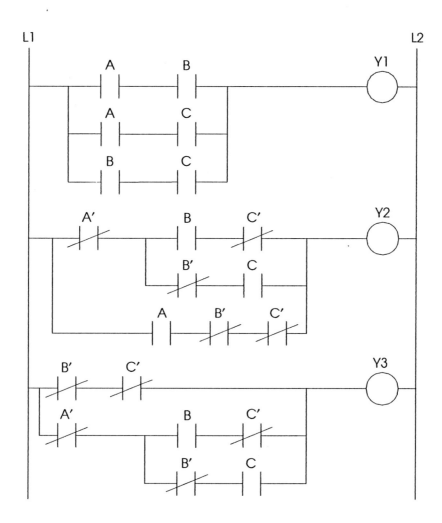

Activity 2.

Step 3.

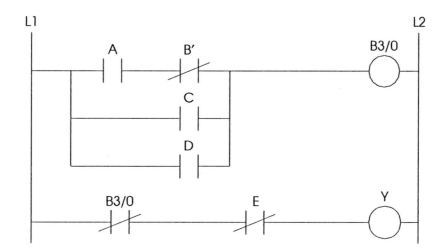

Lab Assignment 14

Activity 1.

Step 1. CR = A + B′ + C + DE

    M + CR + E′

Step 3.

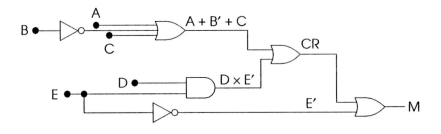

Activity 2.

Step 1. Y1 = C′(A′ + B)

    Y2 = Y1′(C + (A ⊕ D)′)

Step 3.

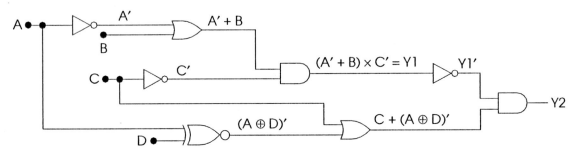

Lab Assignment 15

Activity 1.

Step 1.

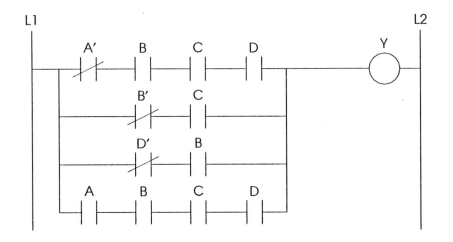

Activity 2.

Step 1.

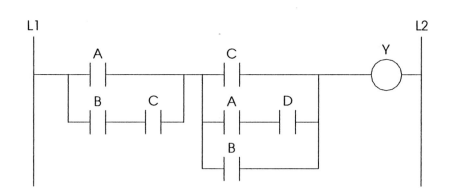

Lab Assignment 16

Activity 1.

Step 3. CR = A(B + CD)E′

$Y = D \times CR'(A + (E \oplus F)')$

Activity 2.

Step 1.

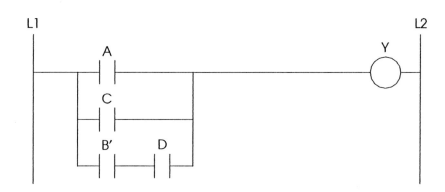

Lab Assignment 17

Activity 1.

Step 1.

SW1	SW2	SW3	SW4	Y
0	0	0	0	0
0	0	0	1	0
0	0	1	0	0
0	0	1	1	0
0	1	0	0	0
0	1	0	1	0
0	1	1	0	0
0	1	1	1	0
1	0	0	0	0
1	0	0	1	0
1	0	1	0	0
1	0	1	1	0
1	1	0	0	0
1	1	0	1	1
1	1	1	0	1
1	1	1	1	1

Step 2. Boolean expression: Y = (SW1)(SW2)(SW3 + SW4)

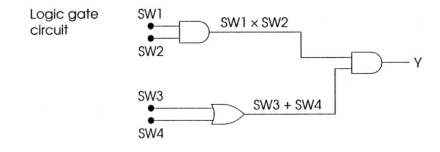

Logic gate circuit

Activity 2.

Step 1.

SW1	SW2	SW3	SW4	Y
0	0	0	0	0
0	0	0	1	0
0	0	1	0	1
0	0	1	1	0
0	1	0	0	1
0	1	0	1	0
0	1	1	0	1
0	1	1	1	0
1	0	0	0	0
1	0	0	1	0
1	0	1	0	0
1	0	1	1	0
1	1	0	0	0
1	1	0	1	0
1	1	1	0	0
1	1	1	1	0

Step 2. Boolean expression: Y = (SW1′)(SW4′)(SW2 + SW3)

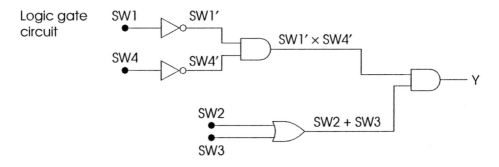

Lab Assignment 18

Activity 1.

Step 1.

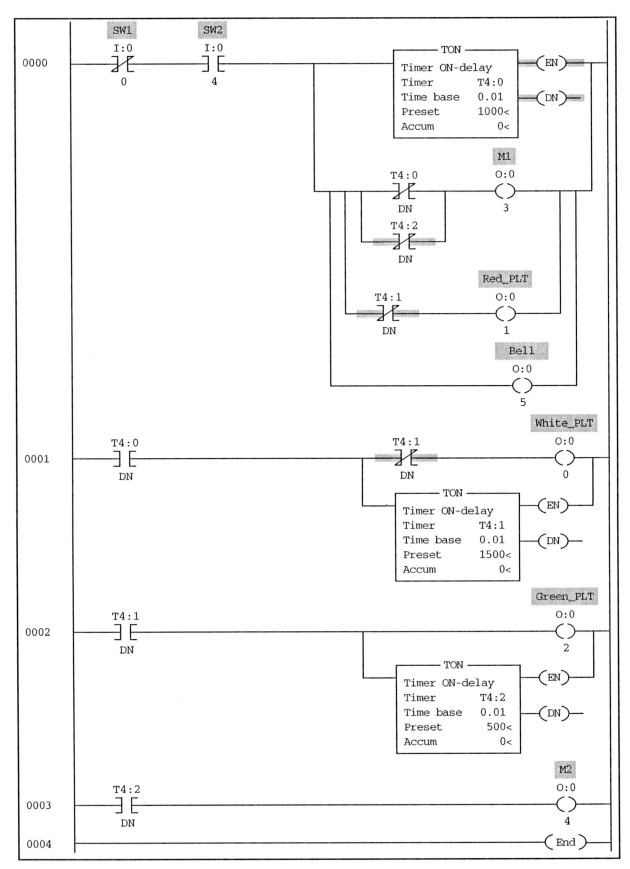

Lab Assignment 19

Activity 1.

Step 1.

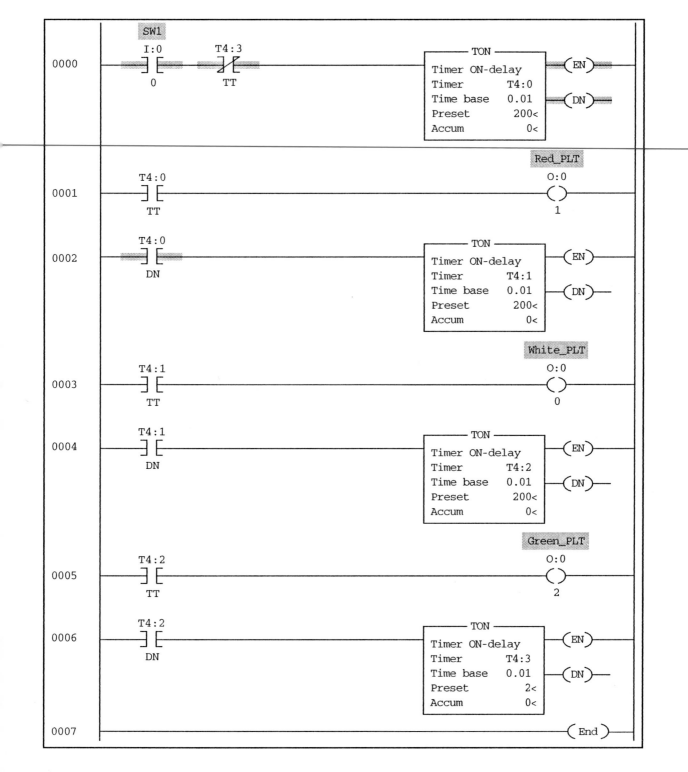

Activity 2.

Step 1.

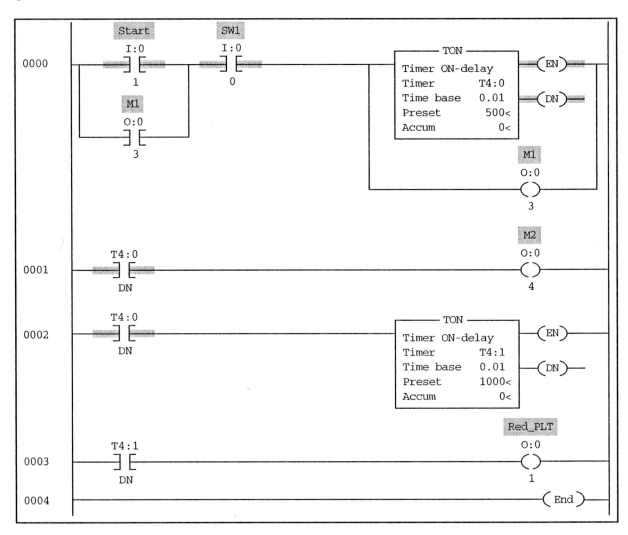

Lab Assignment 20

Activity 1.

Step 1.

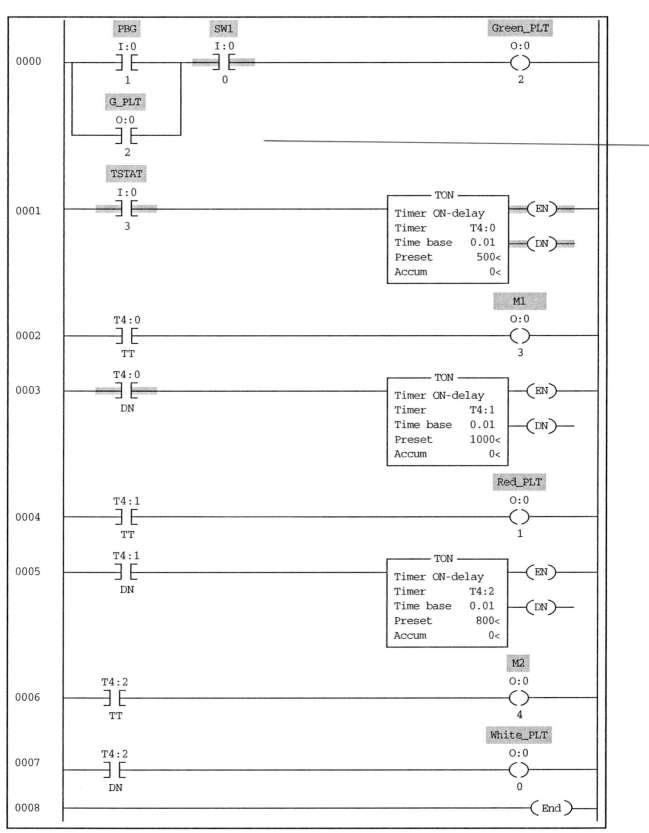

Activity 2.

Step 1.

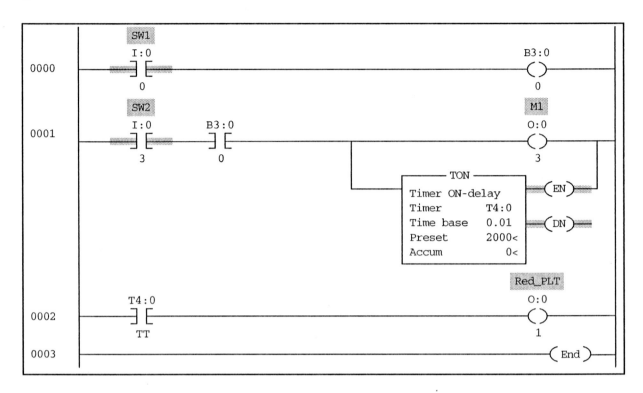

Lab Assignment 21

Activity 1.

Step 1.

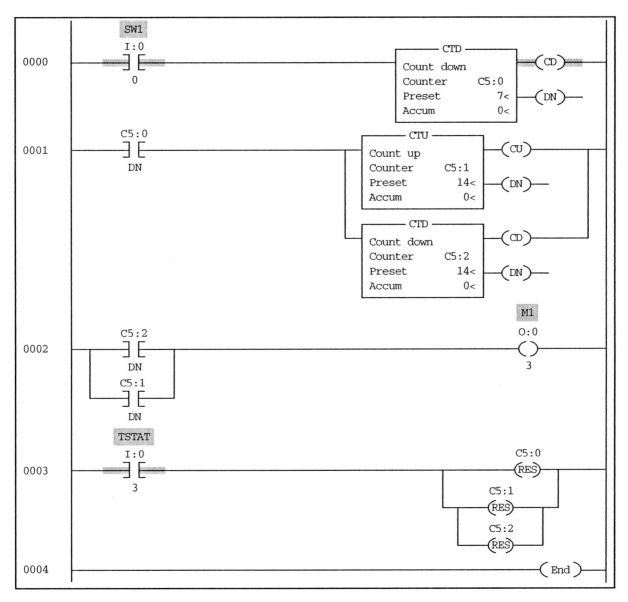

Activity 2.

Step 1.

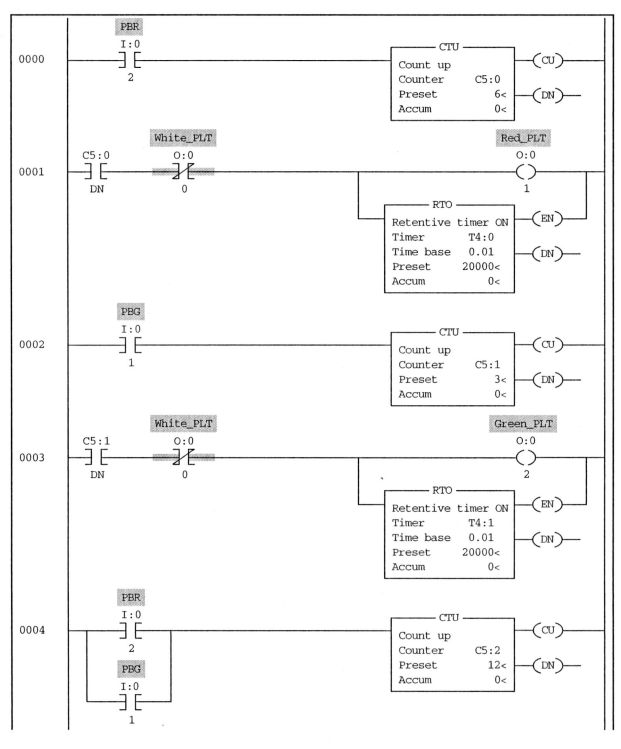

Continued

Step 1. *Continued*

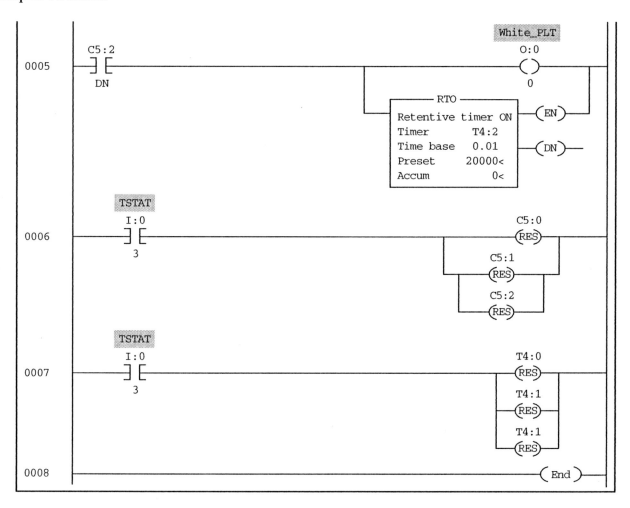

Lab Assignment 22

Activity 1.

Step 1.

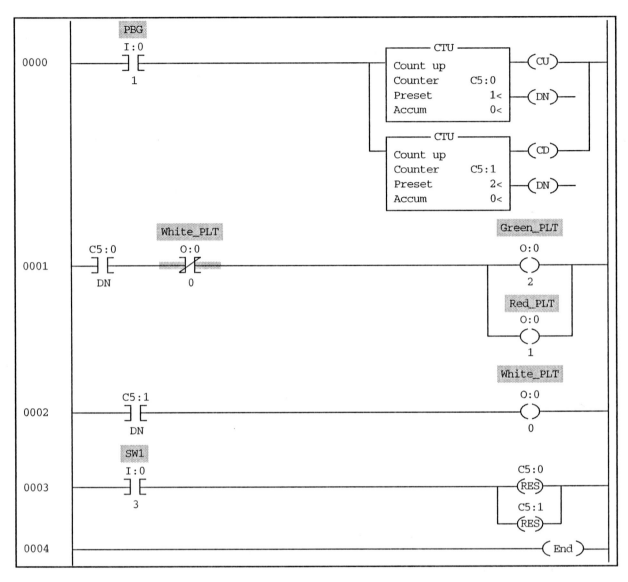

Activity 2.

Step 1.

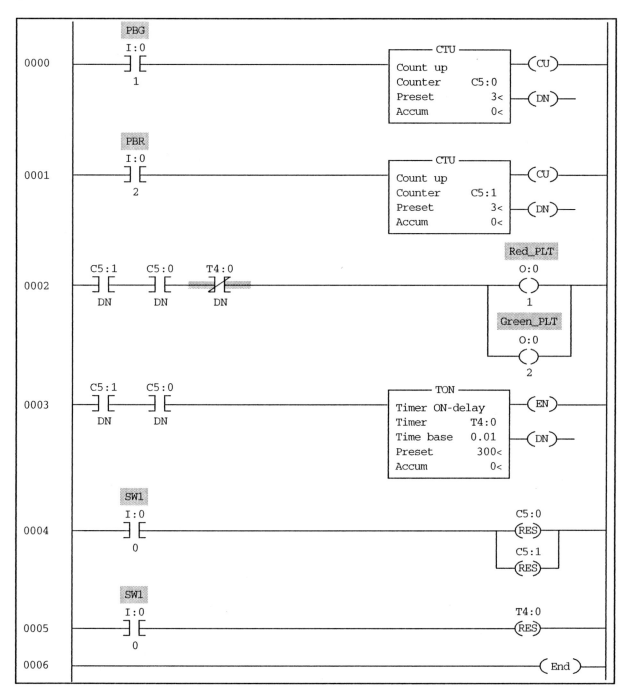

Lab Assignment 23

Activity 1.

Step 1.

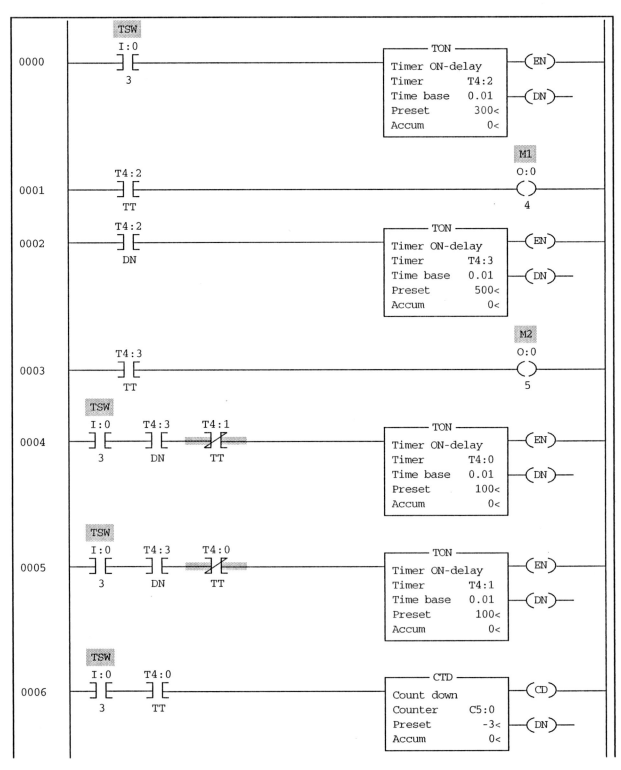

Continued

Step 1. *Continued*

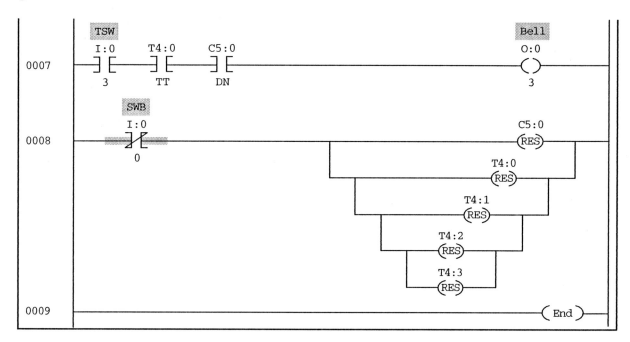

Activity 2.

Step 1.

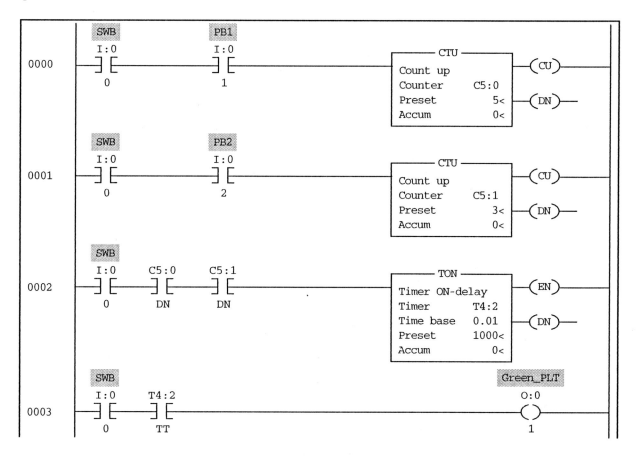

Continued

Step 1. *Continued*

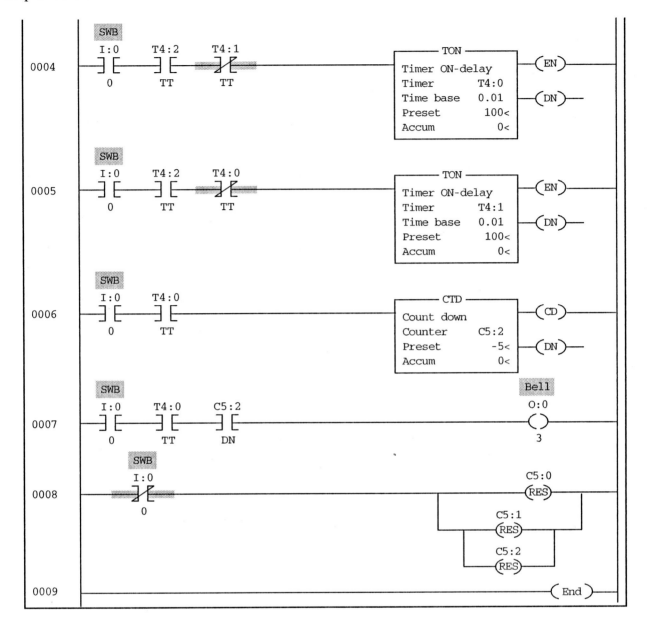

Lab Assignment 24

Activity 1.

Step 1.

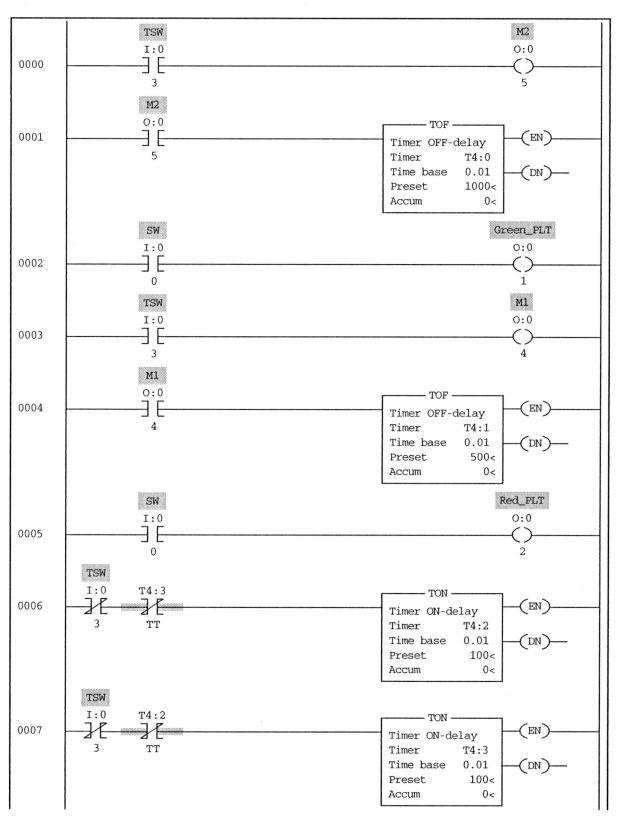

Continued

Step 1. *Continued*

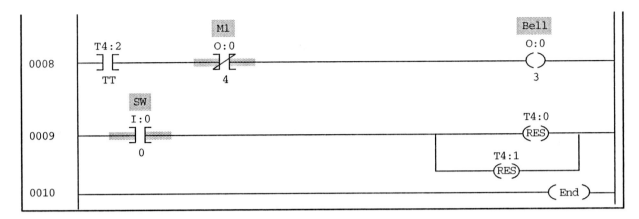

Activity 2.

Step 1.

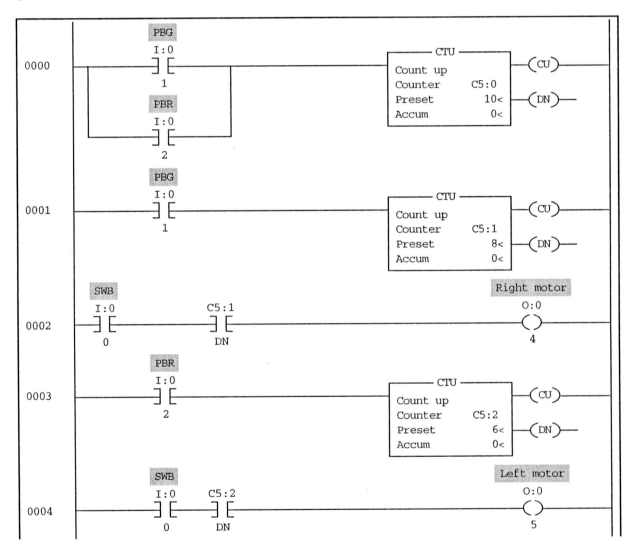

Continued

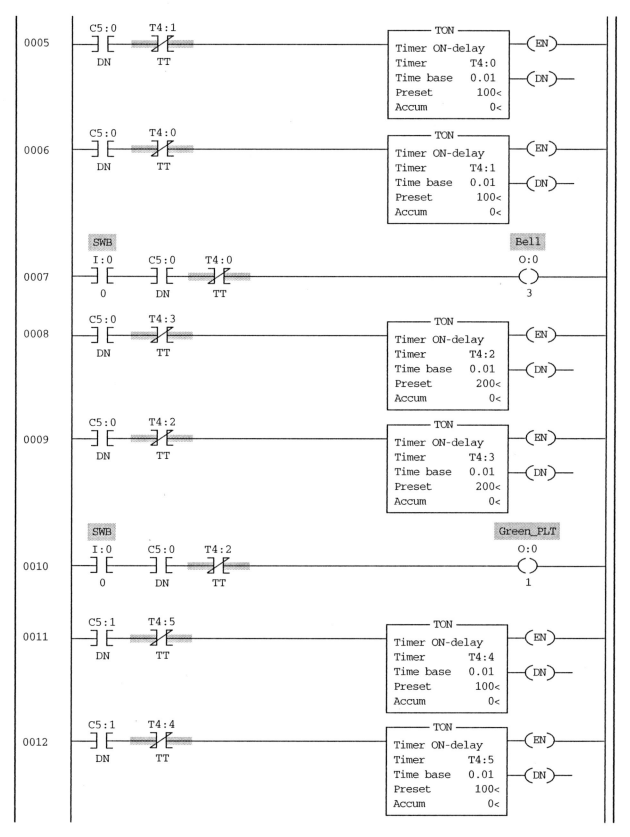

Continued

Step 1. *Continued*

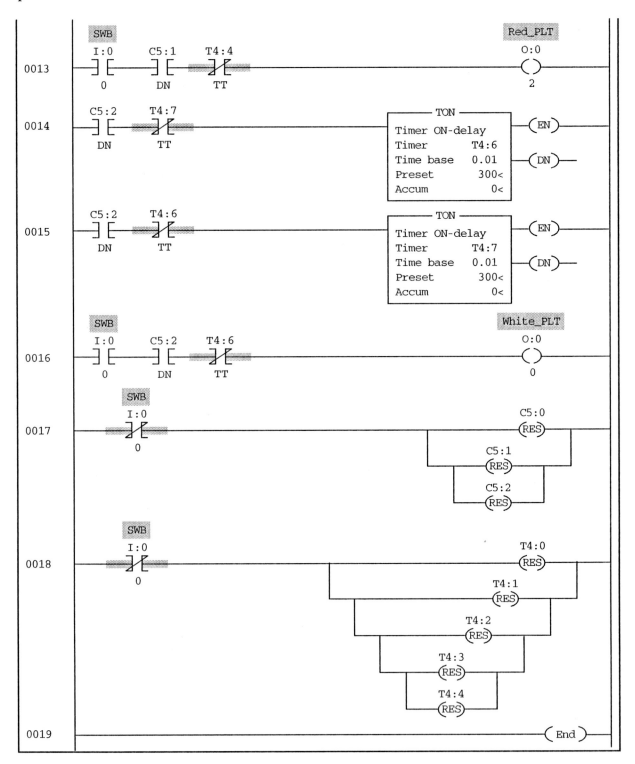

Lab Assignment 25

Activity 1.

Step 1.

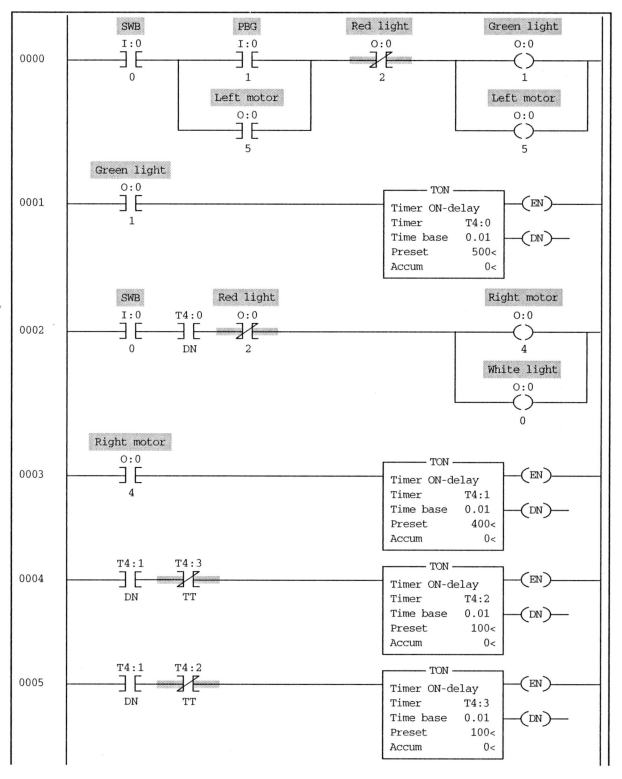

Continued

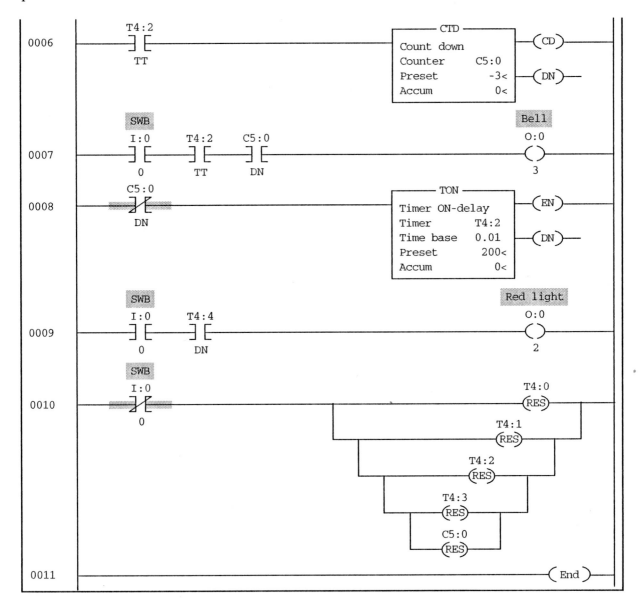

Activity 2.

Step 1.

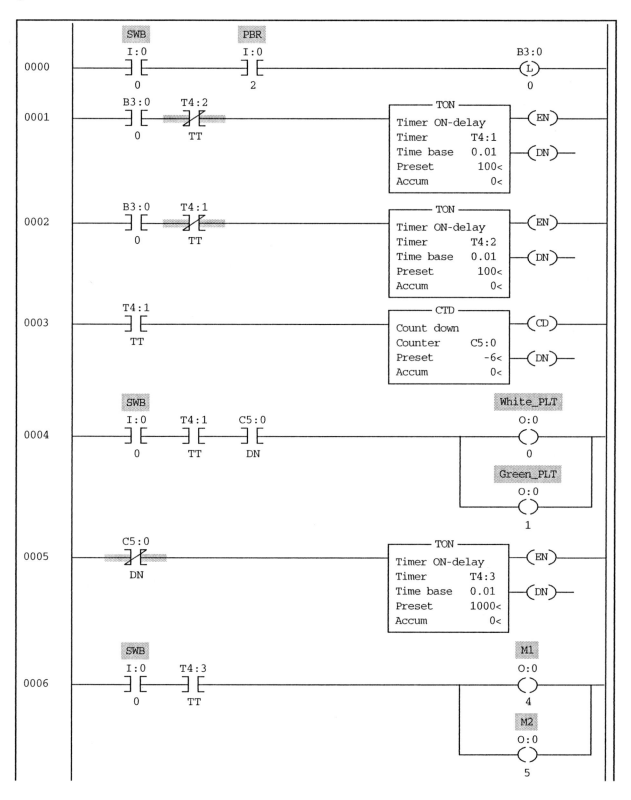

Continued

Step 1. *Continued*

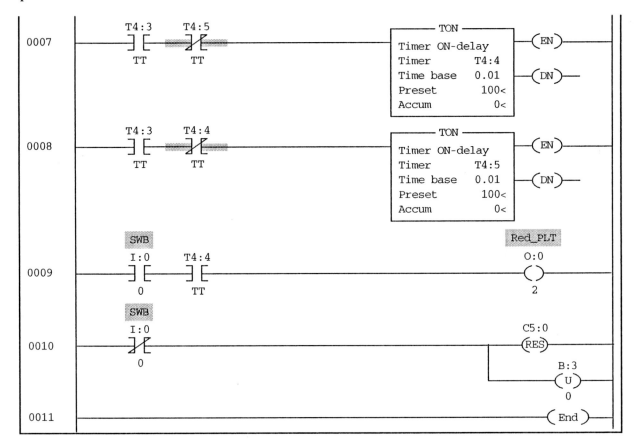

Lab Assignment 26

Activity 1.

Step 1.

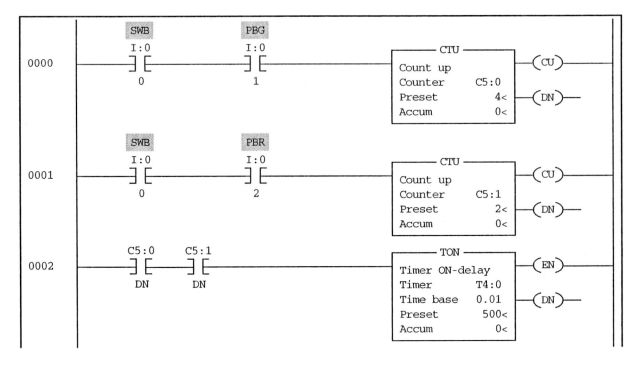

Continued

Step 1. Continued

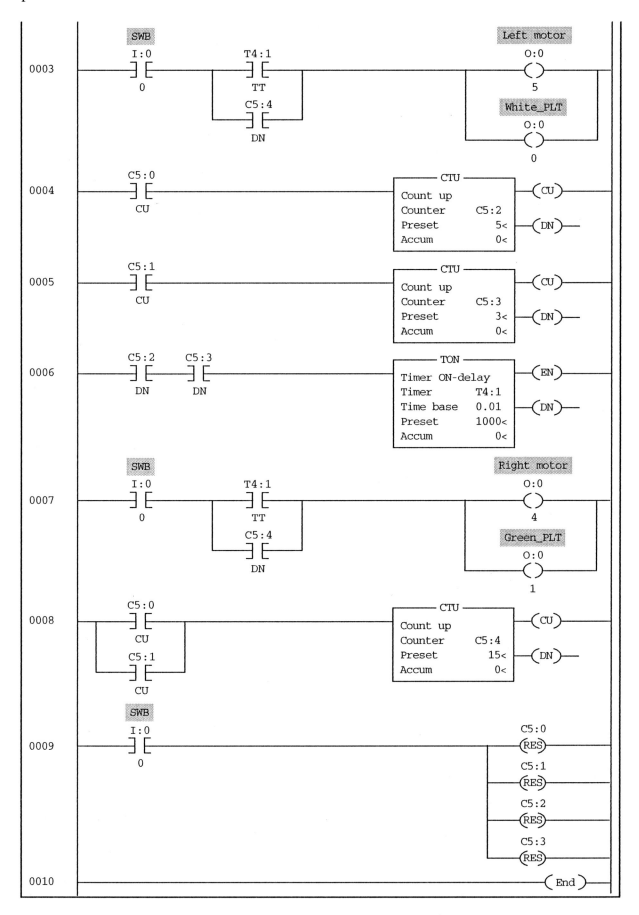

Activity 2.

Step 1.

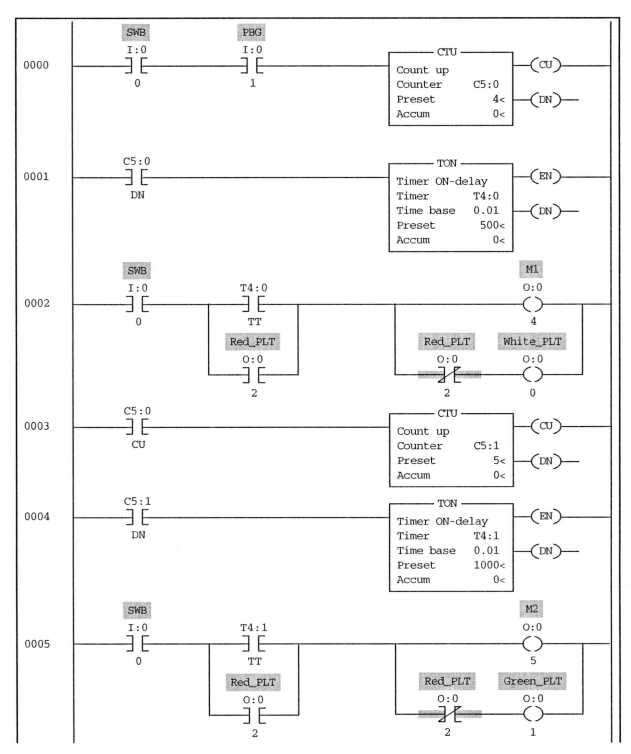

Continued

Step 1. *Continued*

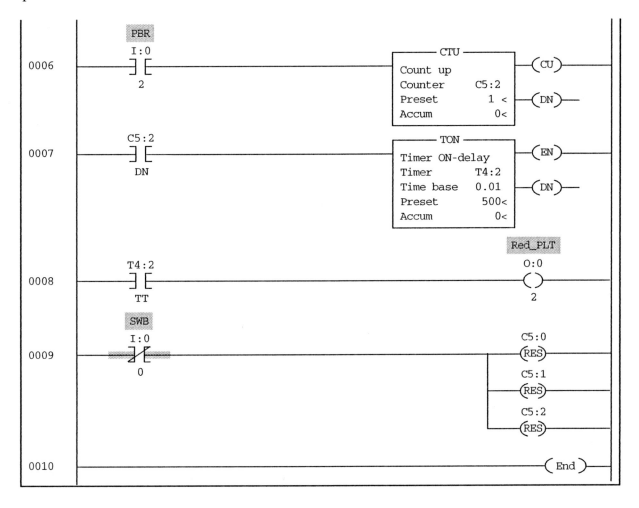

Lab Assignment 27

Activity 1.

Step 1.

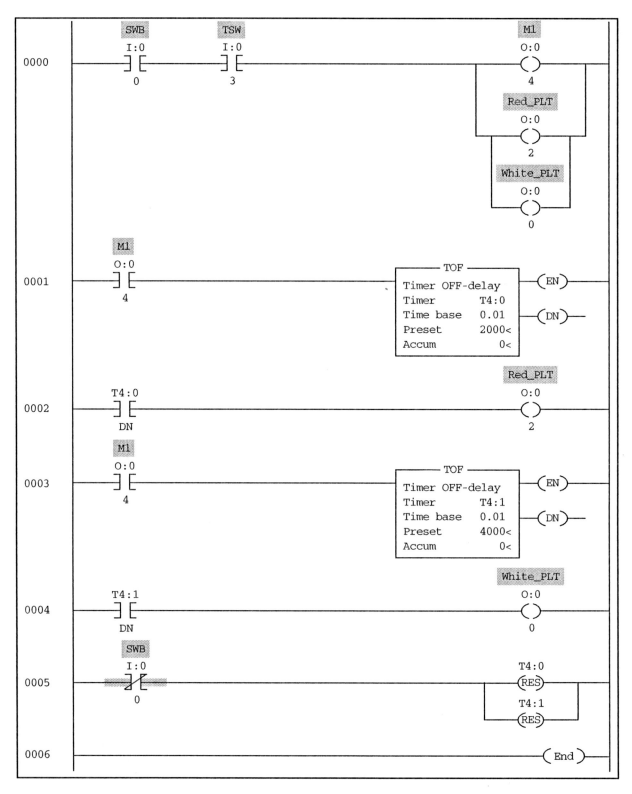

Activity 2.

Step 1.

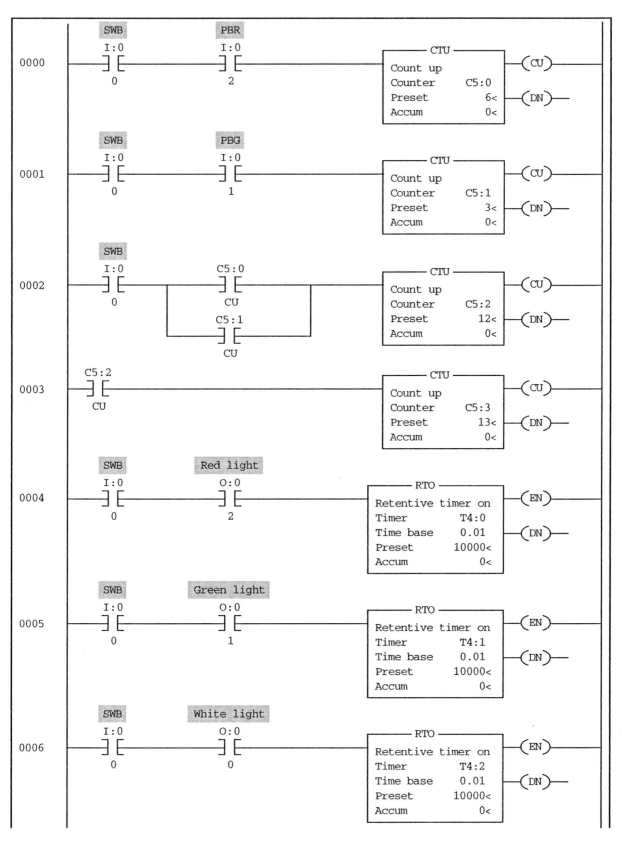

Continued

Step 1. *Continued*

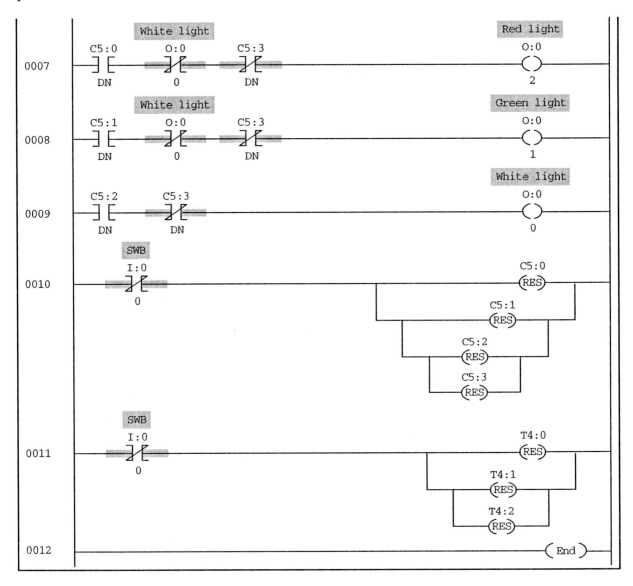

Lab Assignment 28

Activity 1.

Step 1.

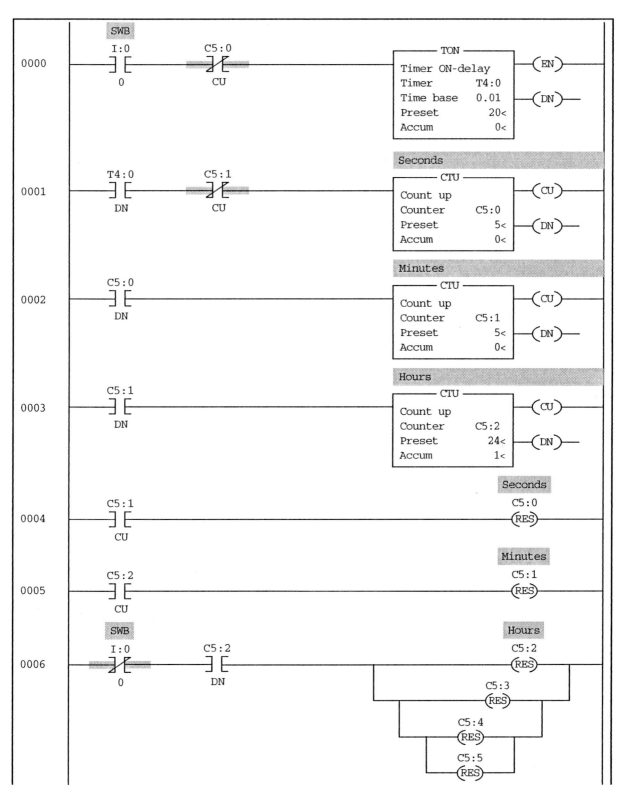

Continued

Step 1. *Continued*

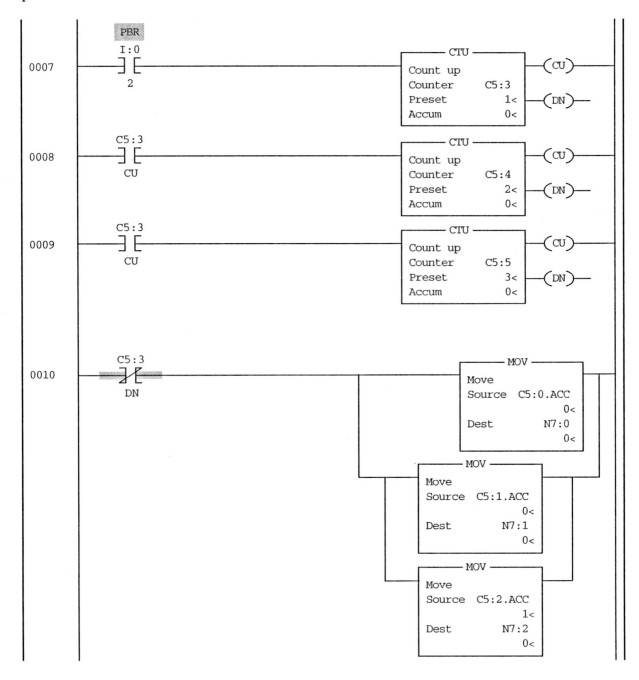

Continued

Step 1. *Continued*

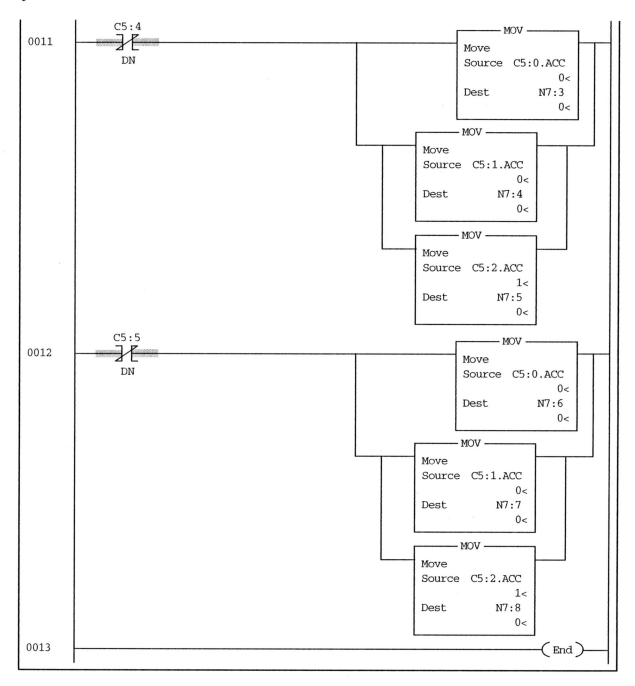

Lab Assignment 29

Activity 1.

The following are possible examples. Student responses may vary.

Addition:

The add function calculates the sum of two operands.

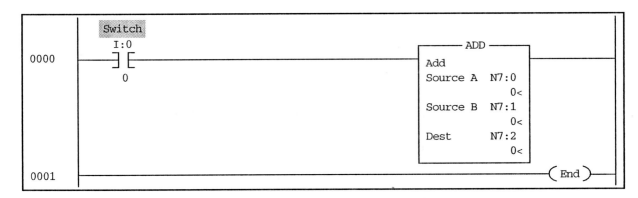

Subtraction:

The subtract function calculates the difference between two sources.

Multiplication:

The multiply function calculates the product of two sources.

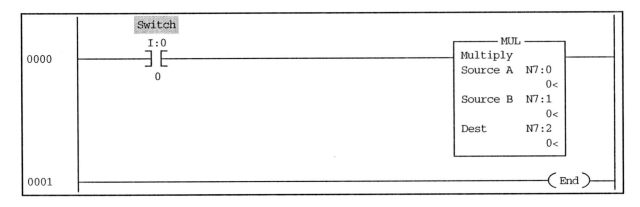

Division:

The divide function calculates the integer value that results from dividing one source by another.

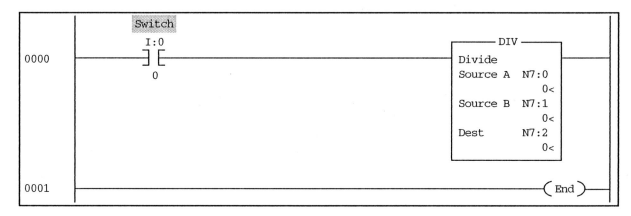

Activity 2.

The following are possible examples. Student responses may vary.

Equal to (EQU):

Checks to see if Source A is equal to Source B.

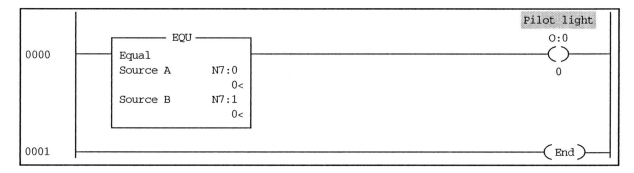

Less than (LES):

Checks to see if the content of Source A is less than the content of Source B.

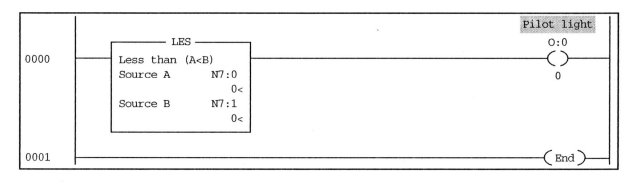

Greater than (GRT):

Checks to see if the content of Source A is greater than the content of Source B.

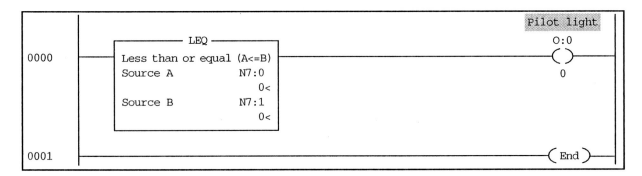

Less than or equal to (LEQ):

Checks to see if the content of Source A is less than or equal to the content of Source B.

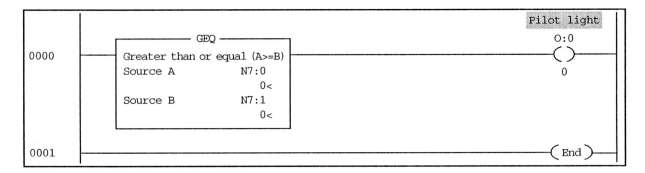

Greater than or equal to (GEQ):

Checks to see if the content of Source A is greater than or equal to the content of Source B.

Activity 3.

Step 1.

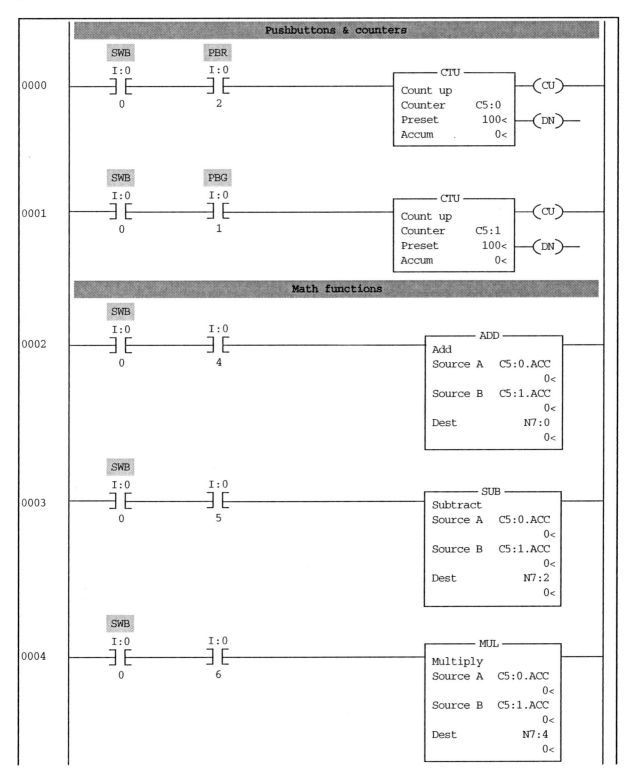

Continued

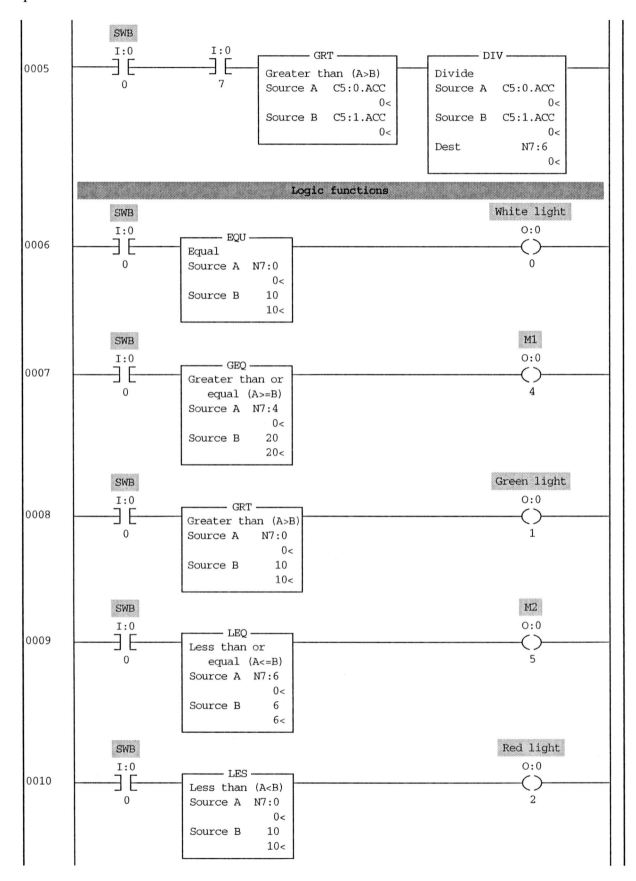

Continued

Step 1. *Continued*

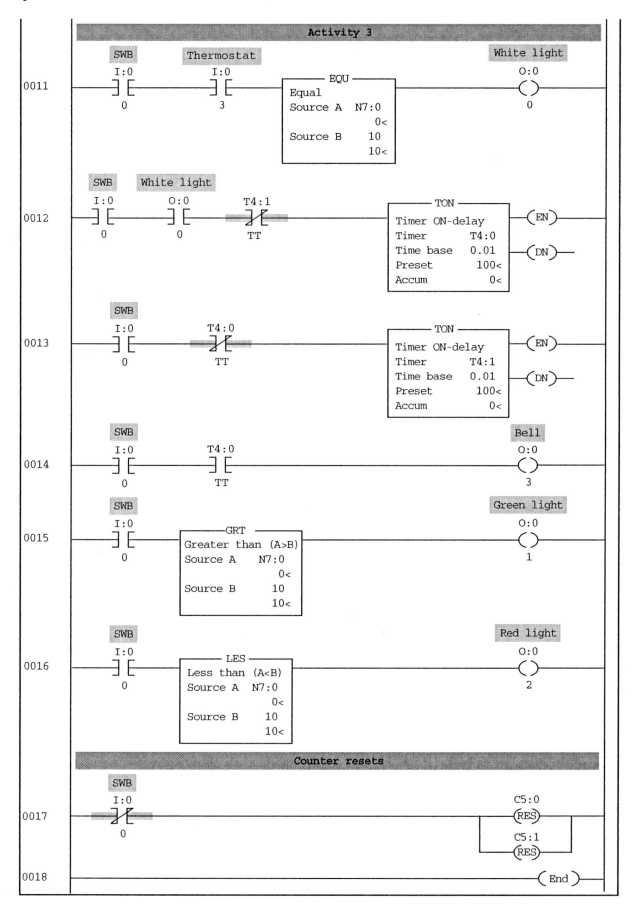

Activity 3

0011 — SWB I:0 0 — Thermostat I:0 3 — EQU / Equal / Source A N7:0 0< / Source B 10 10< — White light O:0 0

0012 — SWB I:0 0 — White light O:0 0 — T4:1 TT — TON / Timer ON-delay / Timer T4:0 / Time base 0.01 / Preset 100< / Accum 0< — (EN) (DN)

0013 — SWB I:0 0 — T4:0 TT — TON / Timer ON-delay / Timer T4:1 / Time base 0.01 / Preset 100< / Accum 0< — (EN) (DN)

0014 — SWB I:0 0 — T4:0 TT — Bell O:0 3

0015 — SWB I:0 0 — GRT / Greater than (A>B) / Source A N7:0 0< / Source B 10 10< — Green light O:0 1

0016 — SWB I:0 0 — LES / Less than (A<B) / Source A N7:0 0< / Source B 10 10< — Red light O:0 2

Counter resets

0017 — SWB I:0 0 — C5:0 (RES) / C5:1 (RES)

0018 — (End)

Lab Assignment 30

Activity 1.

Step 1.

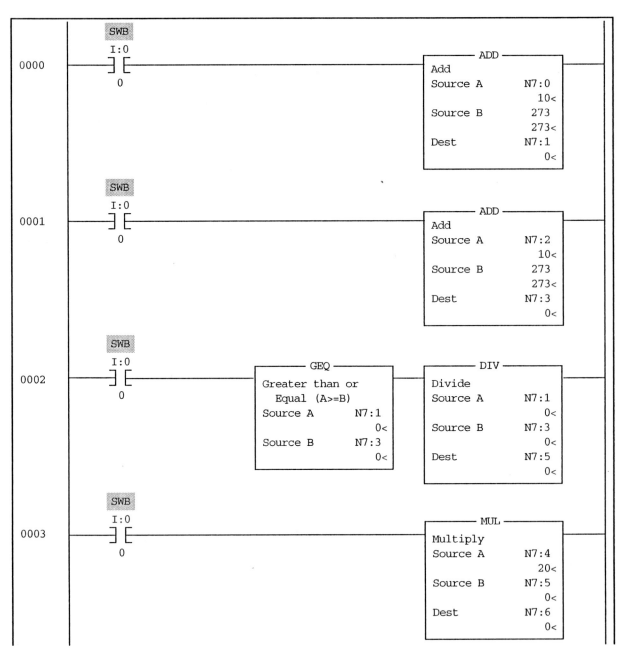

Continued

Step 1. *Continued*

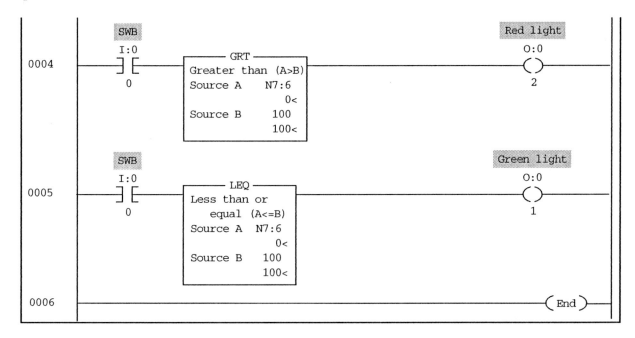

Activity 2.

Step 1.

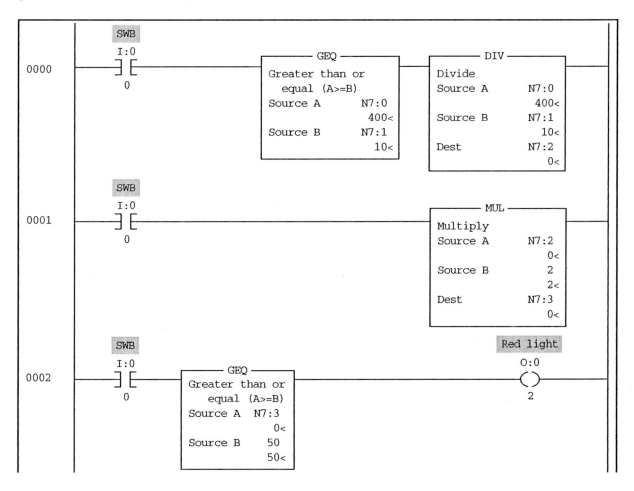

Continued

Step 1. *Continued*

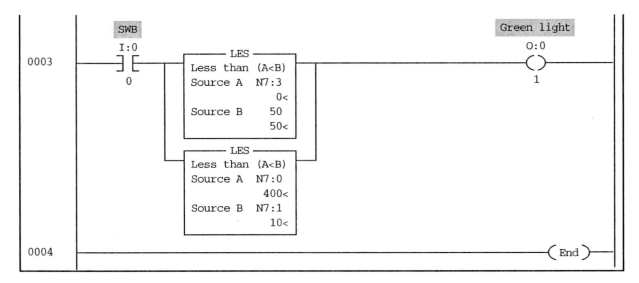

Lab Assignment 31

Activity 1.

Step 1.

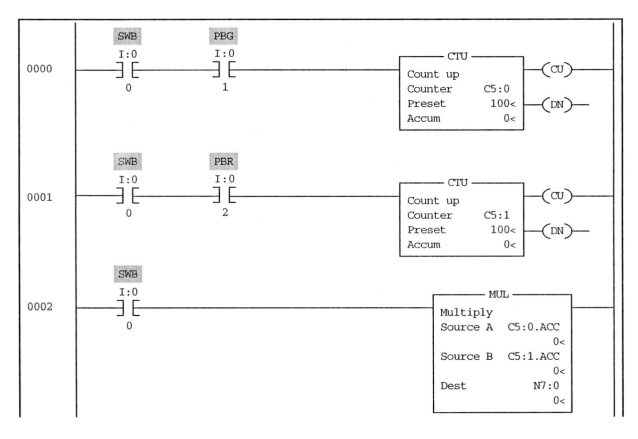

Continued

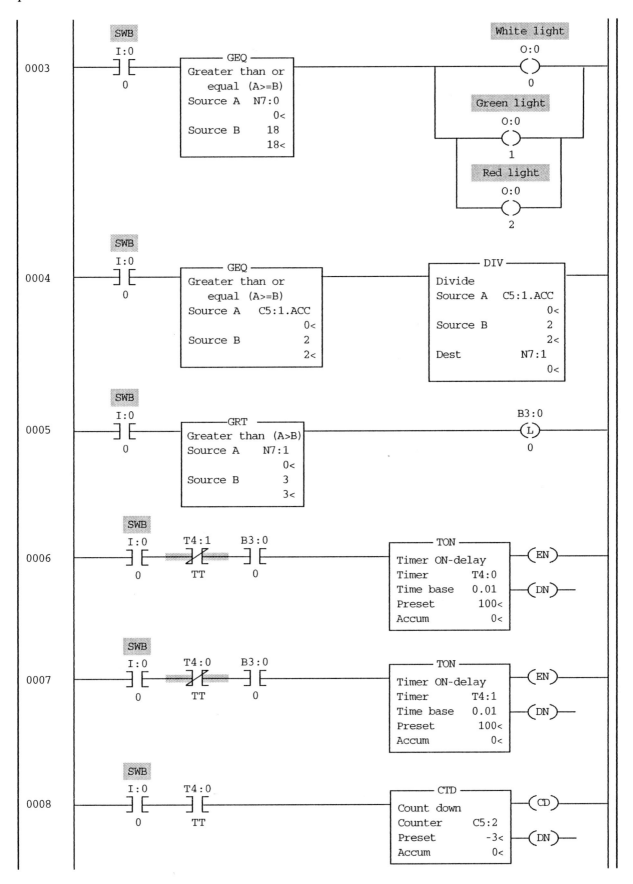

Continued

Step 1. *Continued*

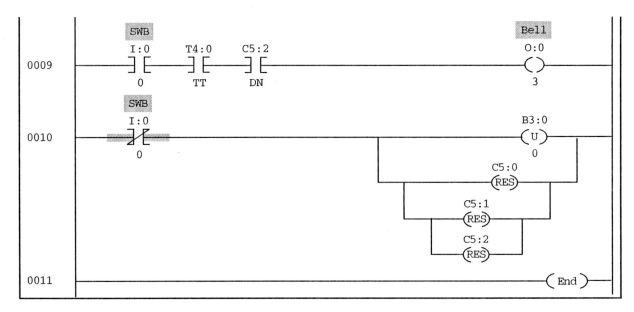

Activity 2.

Step 1.

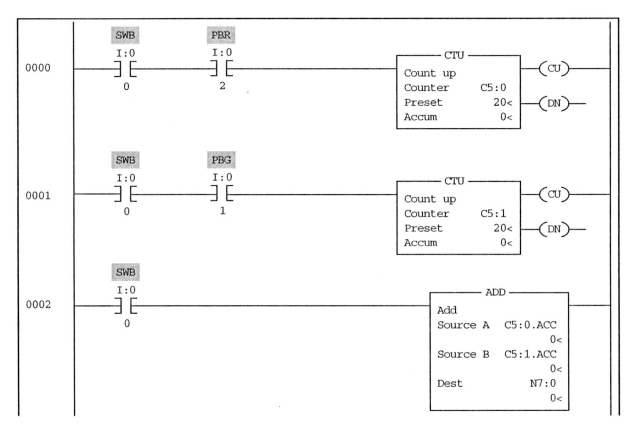

Continued

Step 1. *Continued*

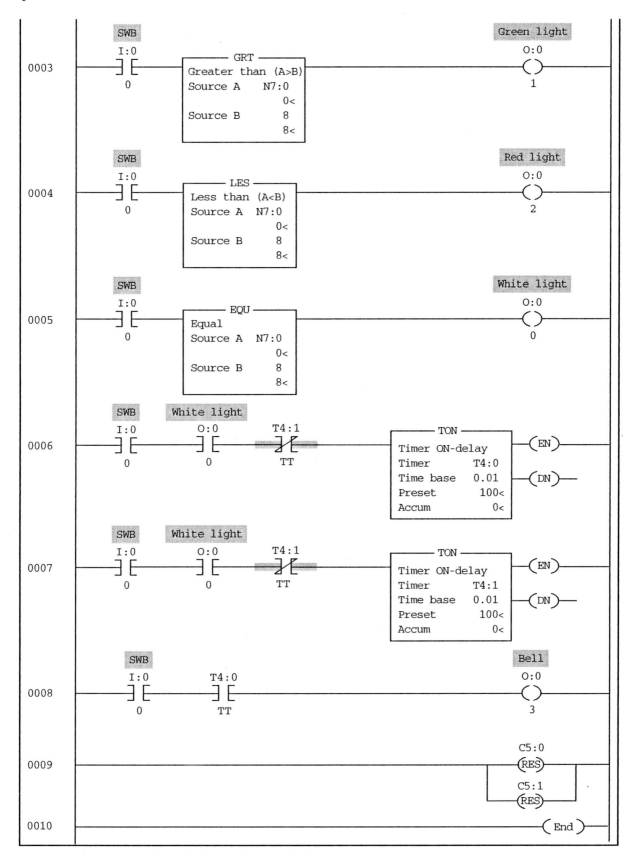

Lab Assignment 32

Activity 1.

Step 1.

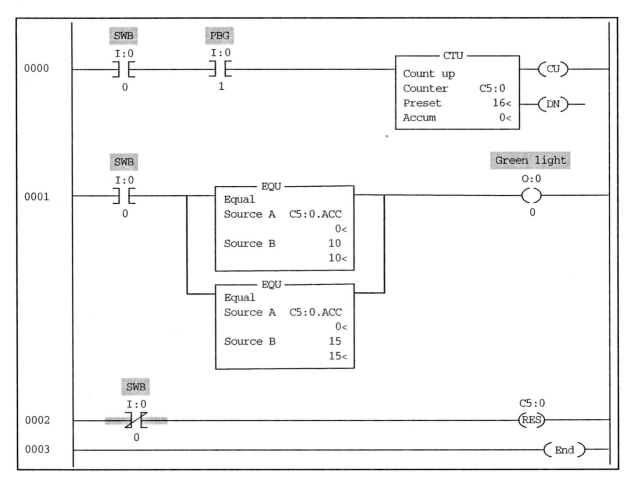

Activity 2.

Step 1.

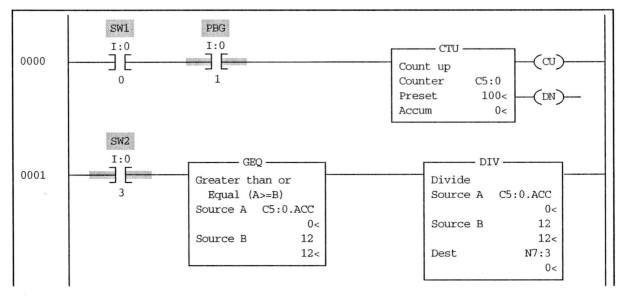

Continued

Step 1. *Continued*

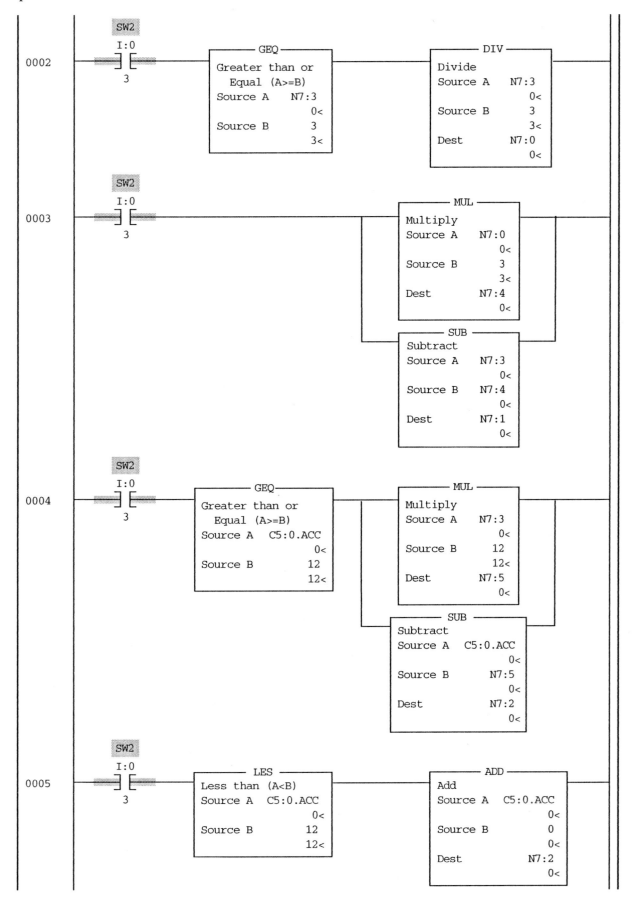

Continued

Step 1. *Continued*

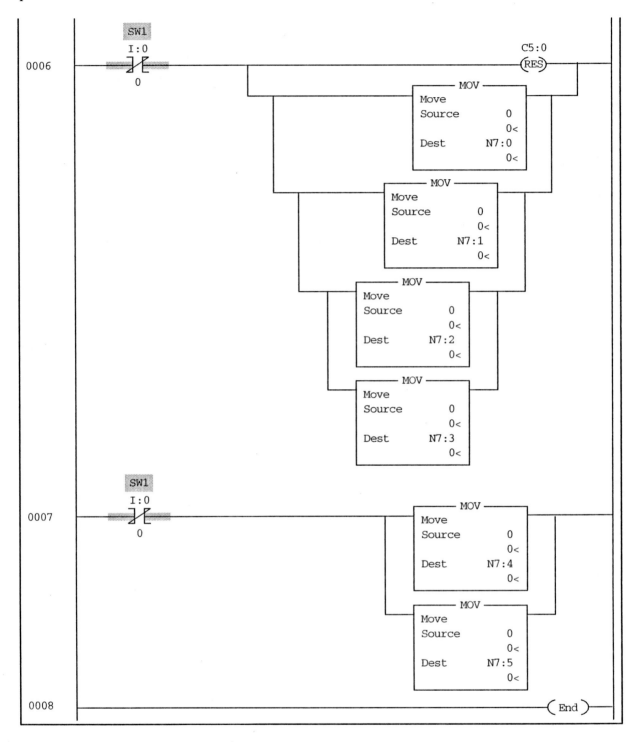

Lab Assignment 33

Activity 1.

Step 1.

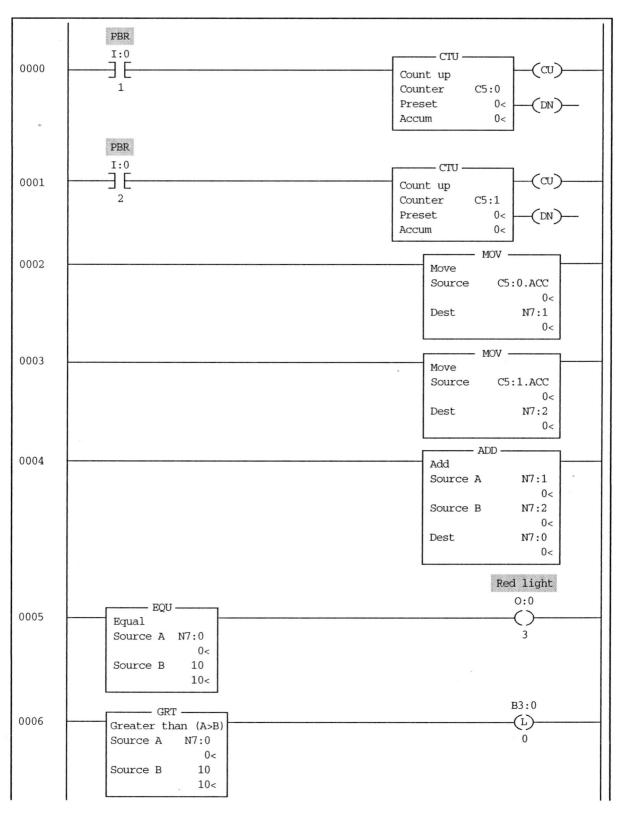

Continued

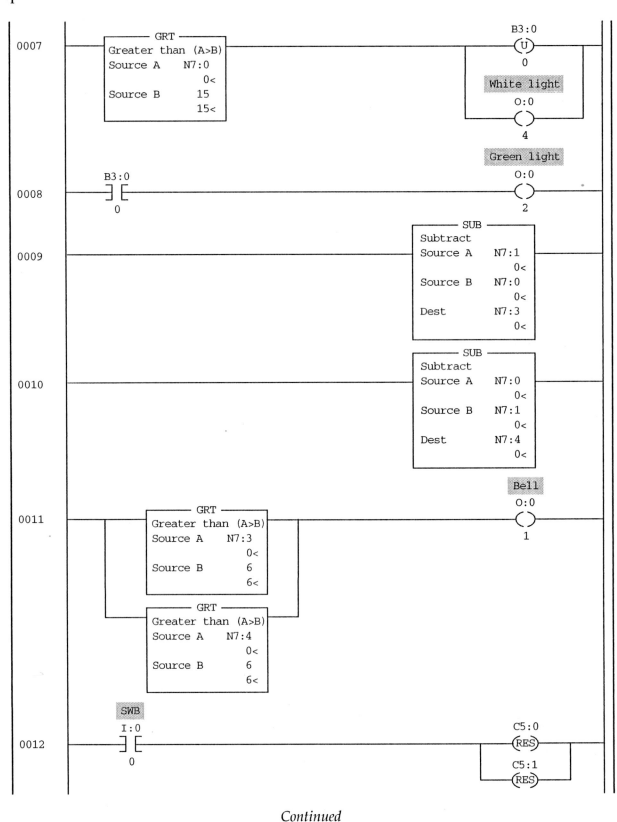

Continued

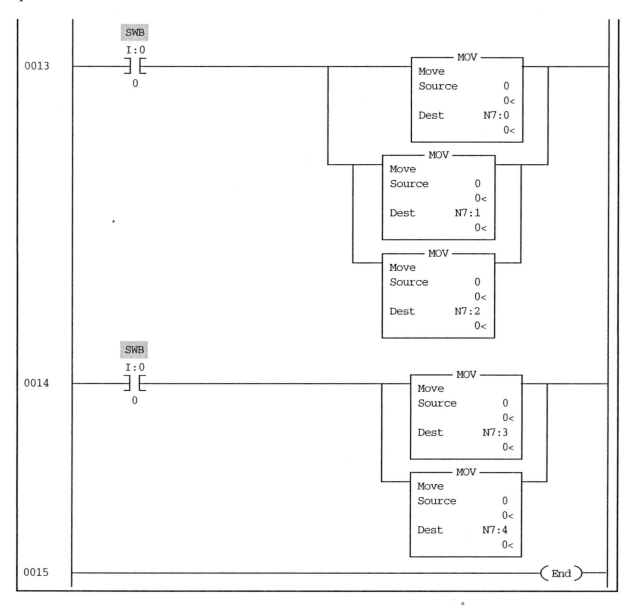

Activity 2.

Step 1.

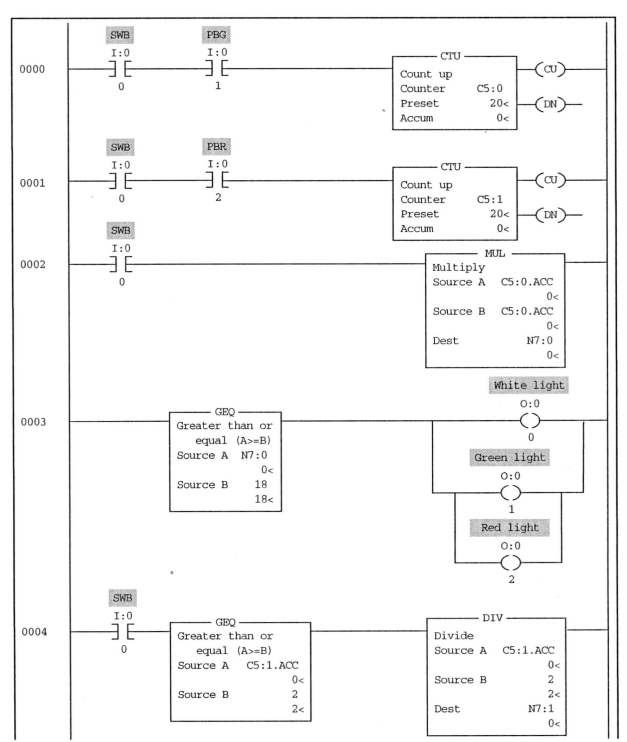

Continued

Step 1. *Continued*

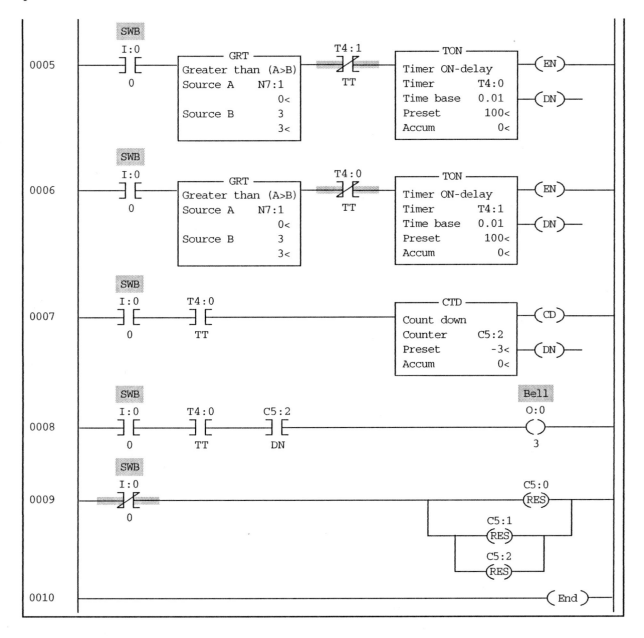

Lab Assignment 34

Activity 1.

Move (MOV): Function used for copying the content of one register into another or for loading a number into a register.

Bit shift left (BSL): When energized, it shifts a bit to the left for every program scan.

Bit shift right (BSR): When energized, it shifts a bit to the right for every program scan.

Step 1.

The following is one possible example. Student responses may vary.

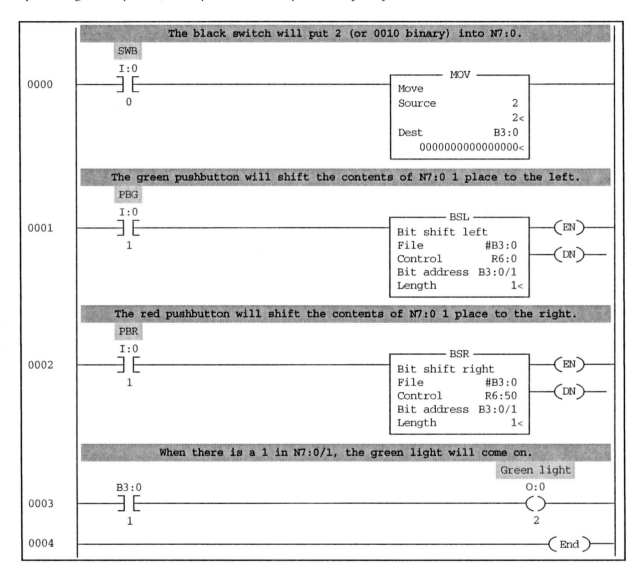

Activity 2.

Step 1.

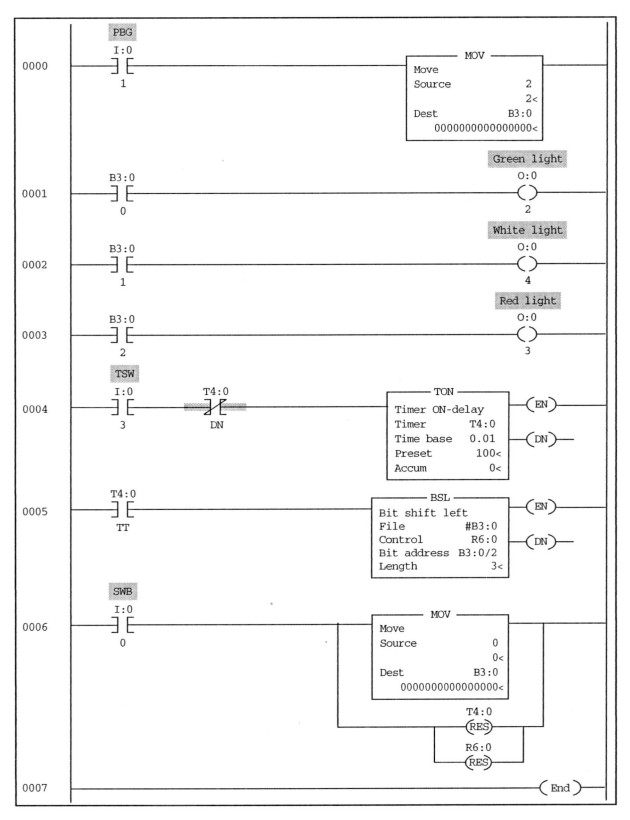

Lab Assignment 35

Activity 1.

Step 1. The jump function is used to skip over rungs. The jump function must be energized to activate it. When the jump function is activated, the instructions between the jump and label function are skipped. The master control reset function is used to stop the operation of the control system during a power outage to the control system. The MCR function must be de-energized to activate it. When the MCR function is activated, the instructions between the MCR functions are turned off.

The following are possible examples. Student responses may vary.

Jump (JMP):

Master control reset (MCR):

Activity 1.

Step 1.

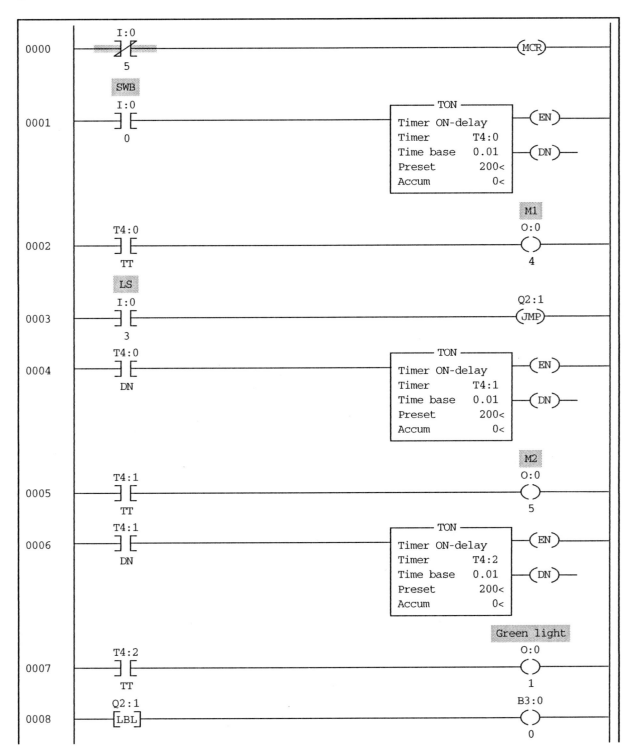

Continued

Step 1. *Continued*

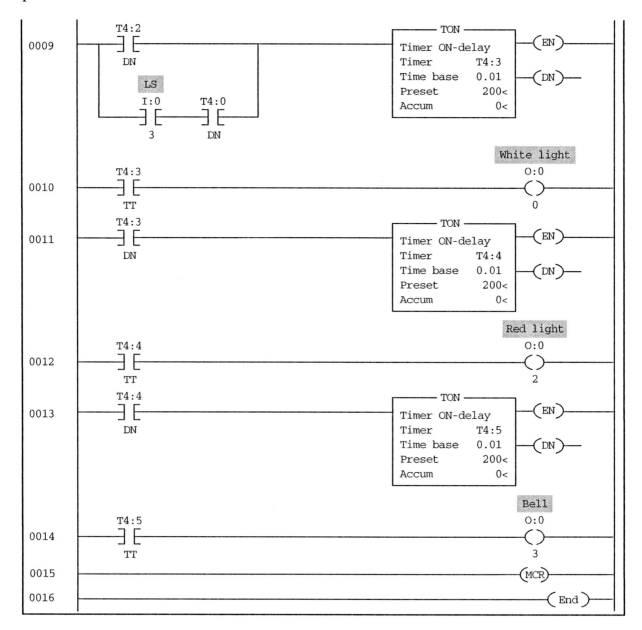

Lab Assignment 36

Activity 1.

Step 1.

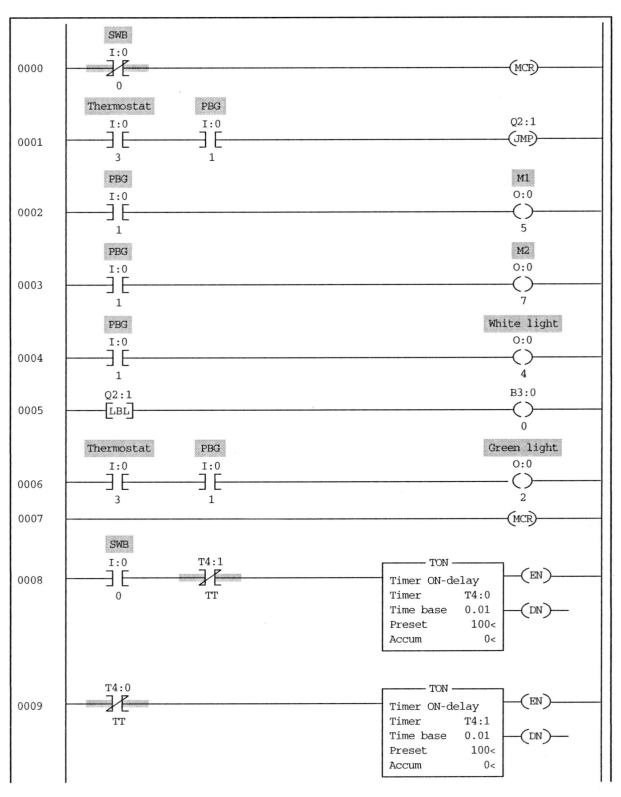

Continued

Step 1. *Continued*

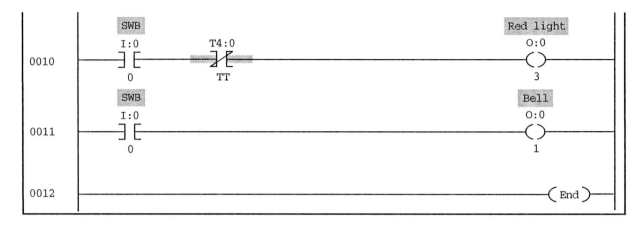

Activity 2.

Step 1.

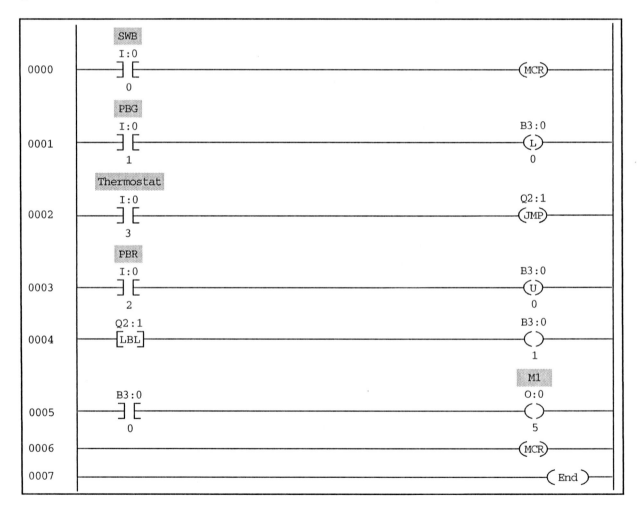

Lab Assignment 37

Activity 1.

Step 1.

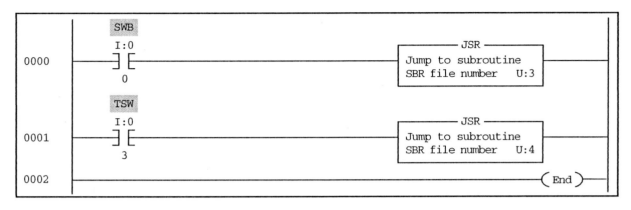

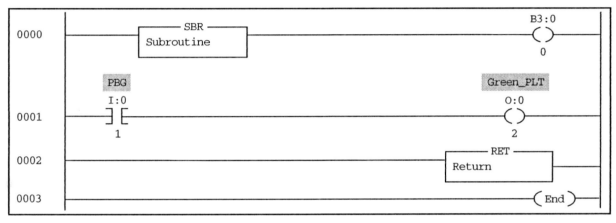

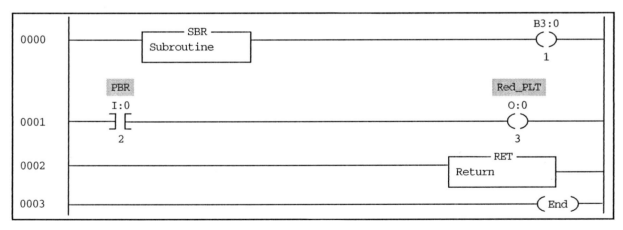

Lab Assignment 38

Activity 1.

Step 1.

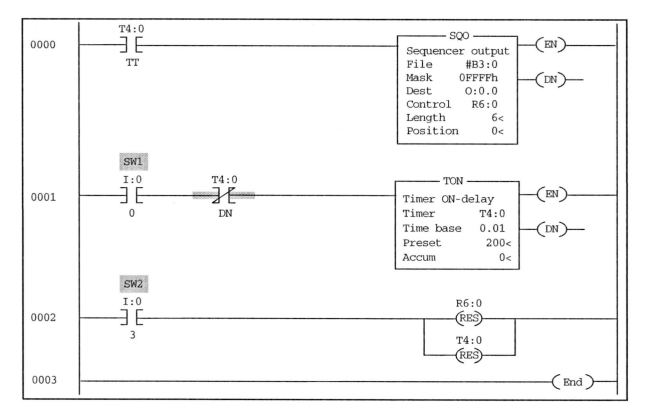

Data files are included for the instructor's benefit. Students are not asked to print the data files with their answers in the laboratory assignments.

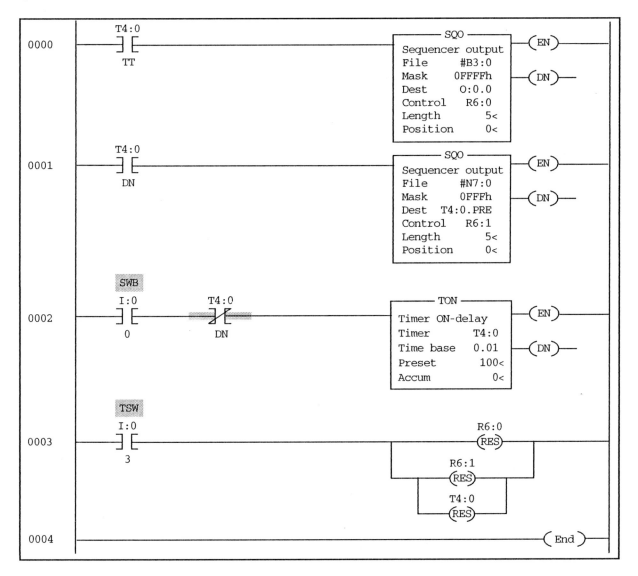

Lab Assignment 39

Activity 1.

Step 1.

Data files are included for the instructor's benefit. Students are not asked to print the data files with their answers in the laboratory assignments.

Data File B3 (bin) -- BINARY

Offset	15	14	13	12	11	10	9	8	7	6	5	4	3	2	1	0
B3:0	0	0	0	0	0	0	0	0	0	0	0	0	0	0	0	0
B3:1	0	0	0	0	0	0	0	0	0	0	1	0	0	0	0	0
B3:2	0	0	0	0	0	0	0	0	0	0	0	0	1	1	0	0
B3:3	0	0	0	0	0	0	0	0	1	0	1	0	0	0	0	0
B3:4	0	0	0	0	0	0	0	0	0	0	0	1	0	0	0	0
B3:5	0	0	0	0	0	0	0	0	0	0	0	0	0	0	0	0

B3:5/0   Radix: Binary

Symbol: Columns: 16

Desc:

B3   Properties   Usage   Help

Data File N7 (dec) -- INTEGER

Offset	0	1	2	3	4	5	6	7	8	9
N7:0	100	200	250	325	150	100				

N7:5   Radix: Decimal

Symbol: Columns: 10

Desc:

N7   Properties   Usage   Help

Lab Assignment 40

Activity 1.

Step 1.

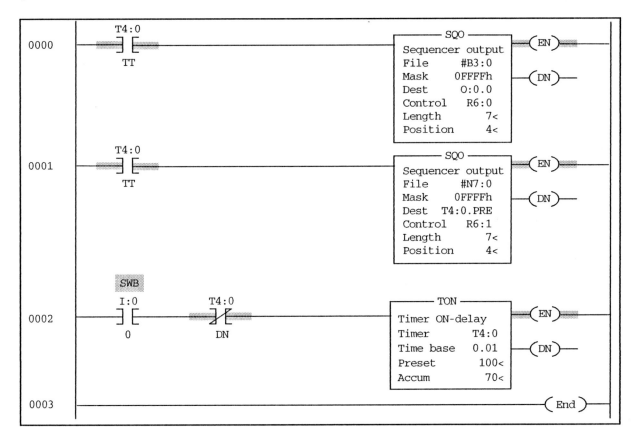

Data files are included for the instructor's benefit. Students are not asked to print the data files with their answers in the laboratory assignments.

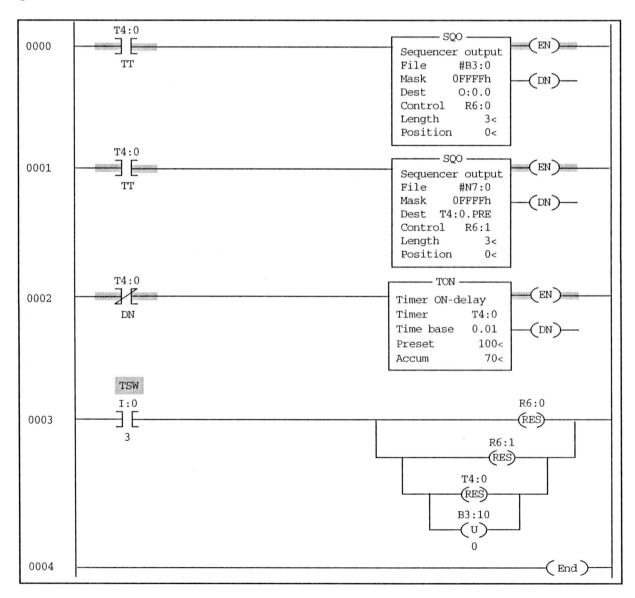

Activity 2.

The diagram for Activity 2 is similar to Activity 1, with different data tables.

Lab Assignment 41

Activity 1.

Step 1.

Data files are included for the instructor's benefit. Students are not asked to print the data files with their answers in the laboratory assignments.

Data File B3 (bin) -- BINARY																	
Offset	15	14	13	12	11	10	9	8	7	6	5	4	3	2	1	0
B3:0	0	0	0	0	0	0	0	0	0	0	0	0	0	0	0	0
B3:1	0	0	0	0	0	0	0	0	0	0	0	0	0	1	0	0
B3:2	0	0	0	0	0	0	0	0	0	0	0	1	0	0	0	0
B3:3	0	0	0	0	0	0	0	0	0	0	0	1	0	0	0	0

B3:0/0   Radix: Binary
Symbol: Columns: 16
Desc:
B3   Properties   Usage   Help

Data File N7 (dec) -- INTEGER										
Offset	0	1	2	3	4	5	6	7	8	9
N7:0	25	3000	500	2000	0	0	0	0		

N7:4   Radix: Decimal
Symbol: Columns: 10
Desc:
N7   Properties   Usage   Help

Activity 2.

Step 1.

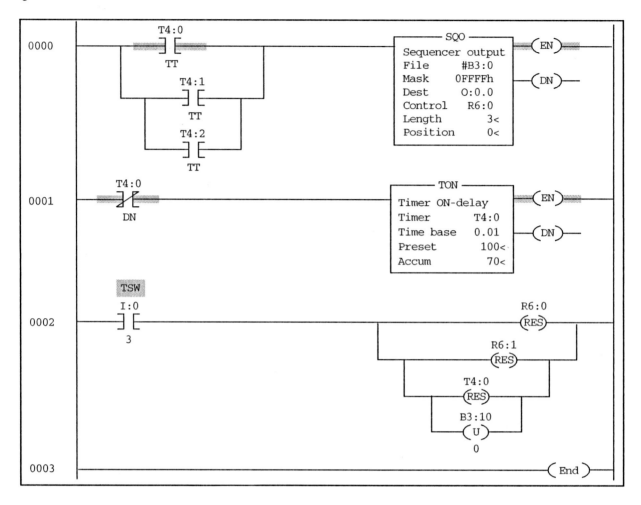

Data files are included for the instructor's benefit. Students are not asked to print the data files with their answers in the laboratory assignments.

Data File B3 (bin) -- BINARY

Offset	15	14	13	12	11	10	9	8	7	6	5	4	3	2	1	0
B3:0	0	0	0	0	0	0	0	0	0	0	0	0	0	0	0	0
B3:1	0	0	0	0	0	0	0	0	0	0	0	0	0	1	0	0
B3:2	0	0	0	0	0	0	0	0	0	0	0	1	0	1	0	0
B3:3	0	0	0	0	0	0	0	0	0	0	1	1	0	1	0	0

B3:0/0   Radix: Binary

Symbol: Columns: 16

Desc:

B3   Properties   Usage   Help

Data File N7 (dec) -- INTEGER

Offset	0	1	2	3	4	5	6	7	8	9
N7:0	25	400	500	1200	800	100	0	0		

N7:6   Radix: Decimal

Symbol: Columns: 10

Desc:

N7   Properties   Usage   Help

Lab Assignment 42

Activity 1.

Step 1.

Master PLC station:

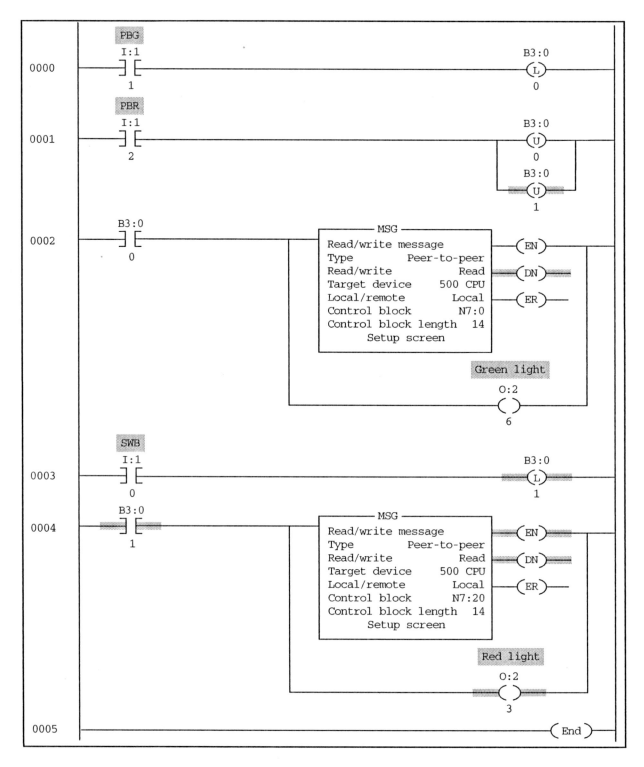

Slave PLC station #1:

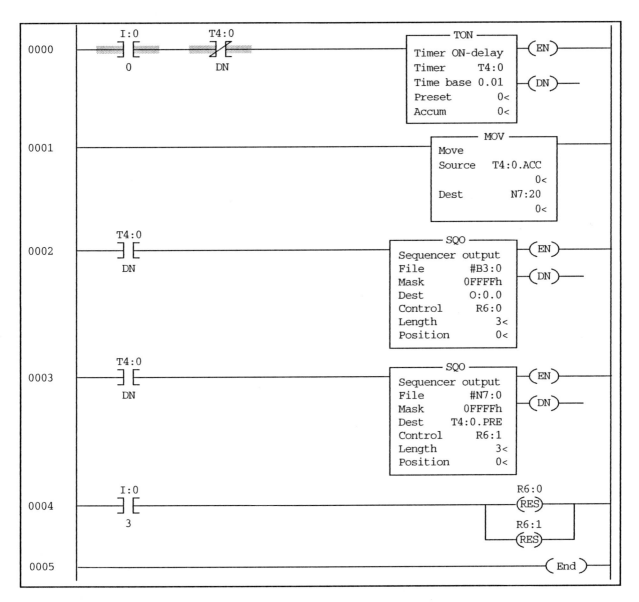

Slave PLC station #2:

See Slave PLC station #1.

Message files.

Control setting for the first read message.

```
MSG - N7:0 : (14 Elements)                                    _□×

 General

 ┌ This Controller ─────────────────────┐  ┌ Control Bits ──────────────────┐
 │ Communication Command : [500CPU Read] │  │   Ignore if timed out (TO):  [0] │
 │       Data Table Address : [N7:15]    │  │        To be retried (NR):  [0] │
 │         Size in Elements : [1]        │  │    Awaiting Execution (EW):  [0] │
 │                Channel : [1]          │  │       Continuous Run (CO):  [0] │
 └───────────────────────────────────────┘  │              Error (ER):  [0] │
                                             │       Message done (DN):  [1] │
 ┌ Target Device ───────────────────────┐  │  Message Transmitting (ST):  [0] │
 │       Message Timeout : [5]           │  │      Message Enabled (EN):  [0] │
 │     Data Table Address : [N7:20]      │  │  Waiting for Queue Space :  [0] │
 │   Local Node Addr (dec): [2]  (octal): [2] └──────────────────────────────────┘
 │       Local / Remote : [Local]        │
 └───────────────────────────────────────┘  ┌ Error ─────────────────────────┐
                                             │   Error Code(Hex):  0          │
                                             └──────────────────────────────────┘
 ┌ Error Description ───────────────────────────────────────────────────────┐
 │     No errors                                                              │
 └───────────────────────────────────────────────────────────────────────────┘
```

Control setting for the second read message.

```
MSG - N7:20 : (14 Elements)                                   _□×

 General

 ┌ This Controller ─────────────────────┐  ┌ Control Bits ──────────────────┐
 │ Communication Command : [500CPU Read] │  │   Ignore if timed out (TO):  [0] │
 │       Data Table Address : [N7:35]    │  │        To be retried (NR):  [0] │
 │         Size in Elements : [1]        │  │    Awaiting Execution (EW):  [0] │
 │                Channel : [1]          │  │       Continuous Run (CO):  [0] │
 └───────────────────────────────────────┘  │              Error (ER):  [0] │
                                             │       Message done (DN):  [1] │
 ┌ Target Device ───────────────────────┐  │  Message Transmitting (ST):  [0] │
 │       Message Timeout : [5]           │  │      Message Enabled (EN):  [0] │
 │     Data Table Address : [N7:20]      │  │  Waiting for Queue Space :  [0] │
 │   Local Node Addr (dec): [1]  (octal): [1] └──────────────────────────────────┘
 │       Local / Remote : [Local]        │
 └───────────────────────────────────────┘  ┌ Error ─────────────────────────┐
                                             │   Error Code(Hex):  0          │
                                             └──────────────────────────────────┘
 ┌ Error Description ───────────────────────────────────────────────────────┐
 │     No errors                                                              │
 └───────────────────────────────────────────────────────────────────────────┘
```

Lab Assignment 43

Activity 1.

Step 1.

Master PLC station:

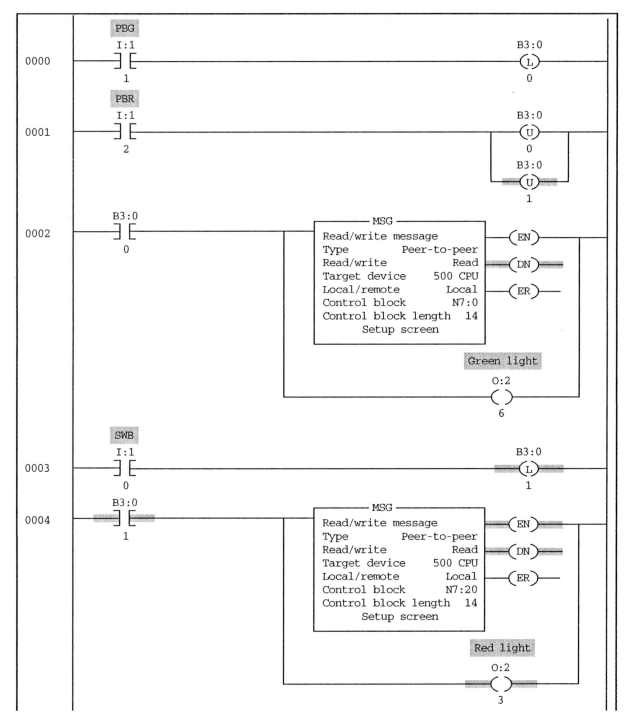

Continued

Step 1. *Continued*

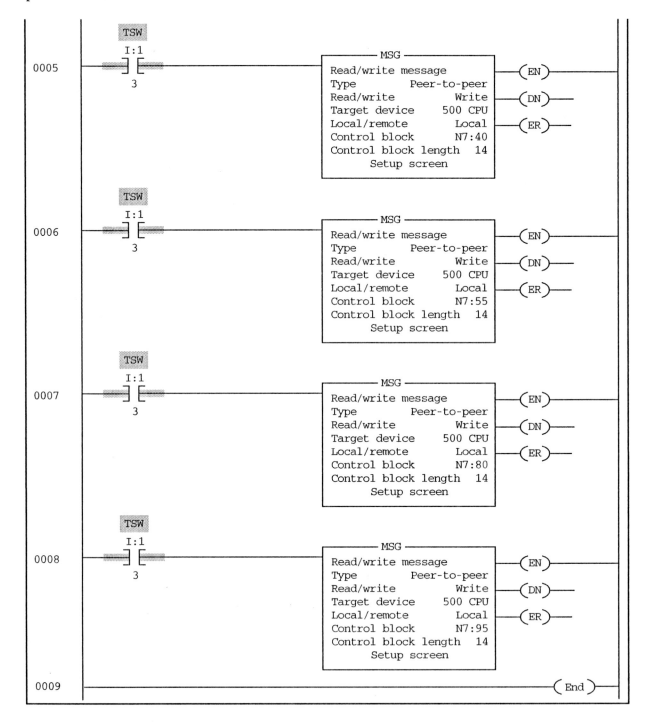

Slave PLC station #1:

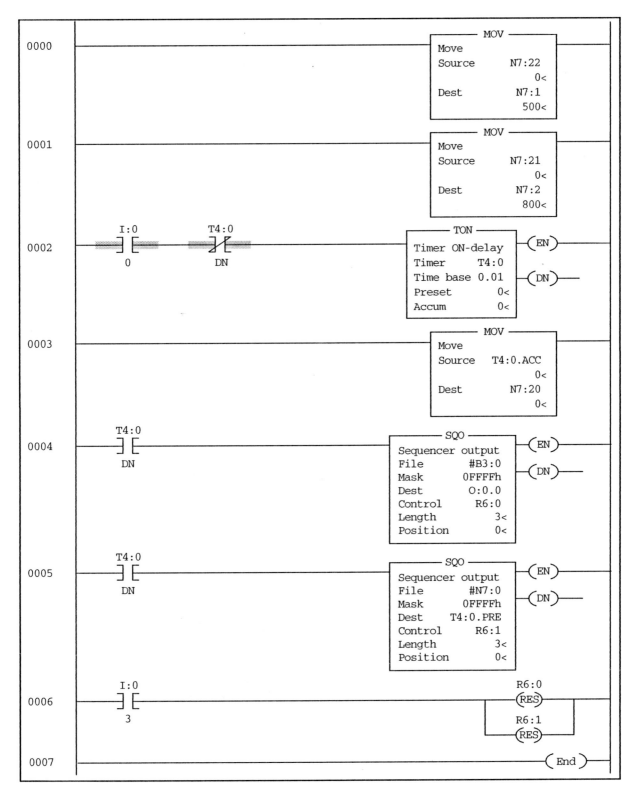

Slave PLC station #2:

See Slave PLC station #1.

Message files.

Control setting for the first read message.

Control setting for the second read message.

Control setting for the first write message.

MSG - N7:40 : (14 Elements)

General

This Controller
Communication Command : 500CPU Write
Data Table Address : N7:36
Size in Elements : 1
Channel: 1

Target Device
Message Timeout : 5
Data Table Address: N7:21
Local Node Addr (dec): 2   (octal): 2
Local / Remote : Local

Control Bits
Ignore if timed out (TO): 0
To be retried (NR): 0
Awaiting Execution (EW): 0
Continuous Run (CO): 0
Error (ER): 0
Message done (DN): 0
Message Transmitting (ST): 0
Message Enabled (EN): 0
Waiting for Queue Space : 0

Error
Error Code(Hex): 0

Error Description
No errors

Control setting for the second write message.

MSG - N7:55 : (14 Elements)

General

This Controller
Communication Command : 500CPU Write
Data Table Address : N7:36
Size in Elements : 1
Channel: 1

Target Device
Message Timeout : 5
Data Table Address: N7:21
Local Node Addr (dec): 1   (octal): 1
Local / Remote : Local

Control Bits
Ignore if timed out (TO): 0
To be retried (NR): 0
Awaiting Execution (EW): 0
Continuous Run (CO): 0
Error (ER): 0
Message done (DN): 0
Message Transmitting (ST): 0
Message Enabled (EN): 0
Waiting for Queue Space : 0

Error
Error Code(Hex): 0

Error Description
No errors

Control setting for the third write message.

Control setting for the fourth write message.

Lab Assignment 44

Activity 1.

Step 1.

Master PLC station:

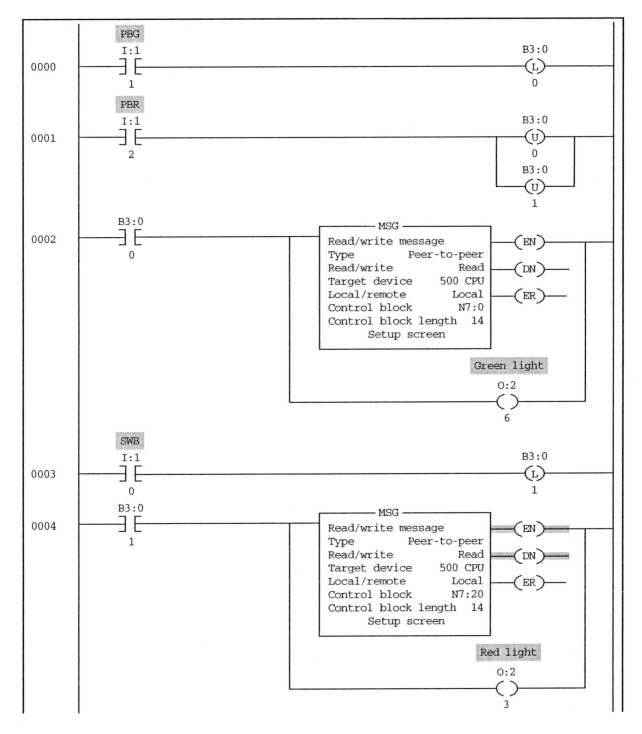

Continued

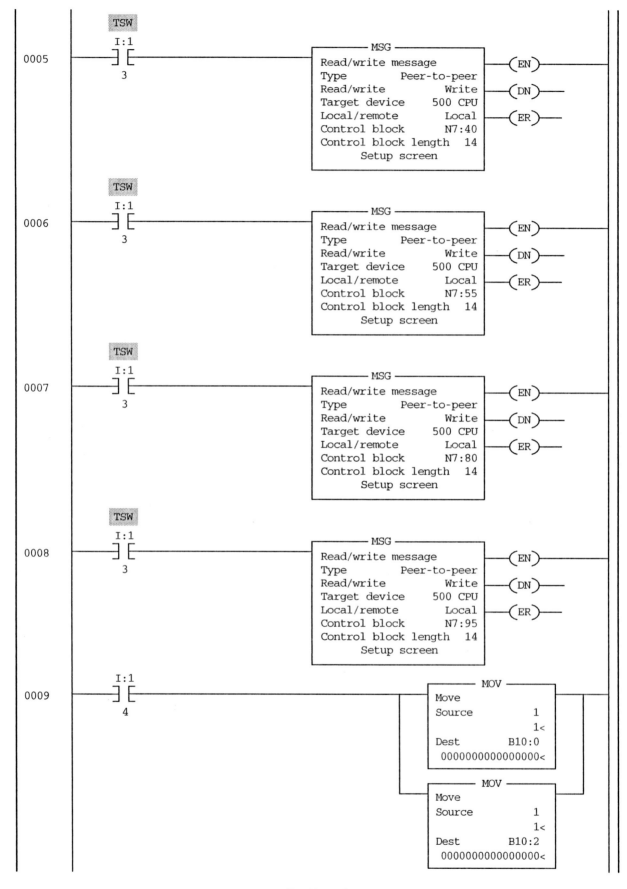

Continued

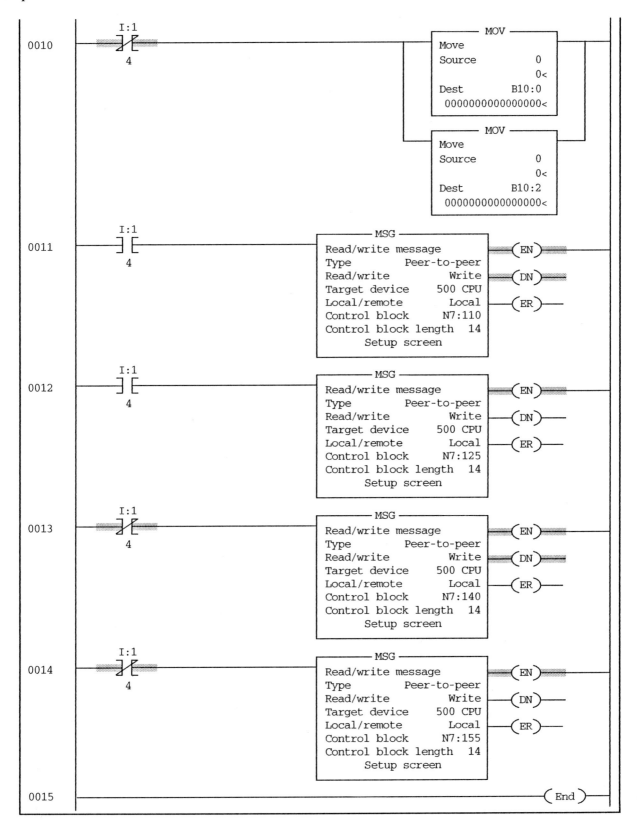

Slave PLC station #1:

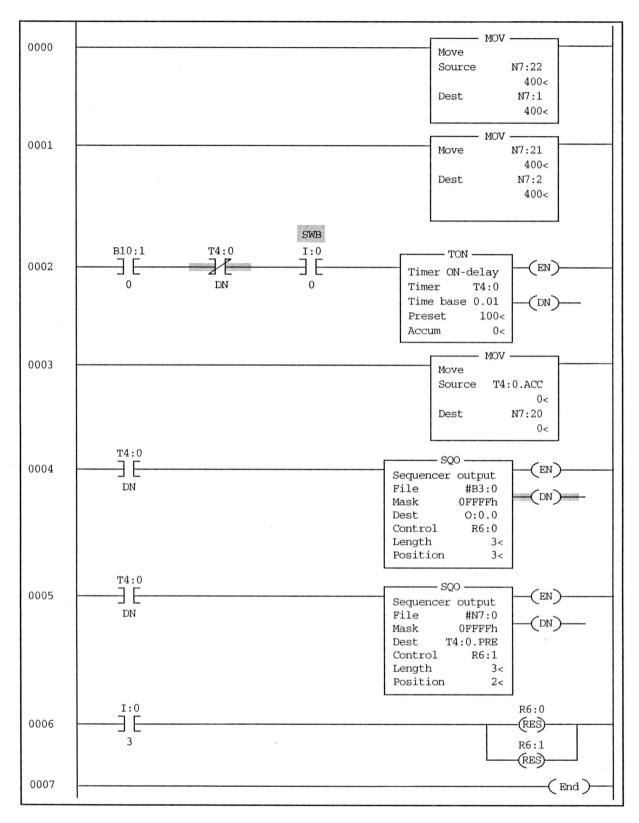

Slave PLC station #2:

See Slave PLC station #1.